KT-362-089

Beyond Reason

4339501003098442772 6012

Beyond Reason

EIGHT GREAT PROBLEMS THAT
REVEAL THE LIMITS OF SCIENCE

A. K. Dewdney

WILEY

John Wiley & Sons, Inc.

HAMPSHIRE COUNTY
LIBRARY

C004035911	
H J	19/08/2004
500	£19.99
0471013986	

Copyright © 2004 by A. K. Dewdney. All rights reserved
Illustrations copyright © 2004 by A. K. Dewdney. All rights reserved

Published by John Wiley & Sons, Inc., Hoboken, New Jersey
Published simultaneously in Canada

Design and production by Navta Associates, Inc.

No part of this publication may be reproduced, stored in a retrieval system, or transmitted in
any form or by any means, electronic, mechanical, photocopying, recording, scanning,
or otherwise, except as permitted under Section 107 or 108 of the 1976 United States
Copyright Act, without either the prior written permission of the Publisher, or authorization
through payment of the appropriate per-copy fee to the Copyright Clearance Center, 222
Rosewood Drive, Danvers, MA 01923, (978) 750-8400, fax (978) 646-8600, or on the web
at www.copyright.com. Requests to the Publisher for permission should be addressed to the
Permissions Department, John Wiley & Sons, Inc., 111 River Street, Hoboken, NJ 07030,
(201) 748-6011, fax (201) 748-6008.

Limit of Liability/Disclaimer of Warranty: While the publisher and the author have used their
best efforts in preparing this book, they make no representations or warranties with respect
to the accuracy or completeness of the contents of this book and specifically disclaim any
implied warranties of merchantability or fitness for a particular purpose. No warranty may be
created or extended by sales representatives or written sales materials. The advice and strate-
gies contained herein may not be suitable for your situation. You should consult with a professional where appropriate. Neither the publisher nor the author shall be liable for any loss
of profit or any other commercial damages, including but not limited to special, incidental,
consequential, or other damages.

For general information about our other products and services, please contact our Customer
Care Department within the United States at (800) 762-2974, outside the United States at
(317) 572-3993 or fax (317) 572-4002.

Wiley also publishes its books in a variety of electronic formats. Some content that appears in
print may not be available in electronic books. For more information about Wiley products,
visit our web site at www.wiley.com.

ISBN 0-471-01398-6

Printed in the United States of America

10 9 8 7 6 5 4 3 2 1

·CONTENTS·

Beyond Reason

Where Reason Cannot Go

WE ALL LIVE WITH LIMITATIONS—some natural, some rule-based. We cannot fly unaided, nor can we beat the ace of spades with the deuce. This book is about the grander limitations that stand like granite walls around our scientific and technological enterprise. "This far and no farther," they seem to say, an ancient prohibition from sacred ground. I locate these barriers "beyond reason" because reason, even though it found them, sees no way around them.

Potential barriers might include a prohibition of time travel, especially into the past. However, I know of no established physical or mathematical theory that prohibits it. There may also be some reason why we can never solve the prime number problem, but I do not know it.

To speak of time travel and prime numbers in one breath expresses the dual nature of this book. Everyone knows what "time travel" means, but few know what a prime number is. Thus I must immediately explain that a prime number is one that cannot be divided evenly by any other number except 1. To solve the prime number problem means to arrive at what mathematicians call a closed-form expression involving a variable n that, when you substitute specific values of n into the expression, yields the nth prime number. I know of no theory that prohibits such a possibility, although the problem may well have no solution.

This book deals with two kinds of science: one inductive, the other deductive. One is vastly more popular than the other, but could not exist without it.

Although no current physical theory appears to prohibit time travel, the discovery of a method is sure to be cloaked in the language of mathematics. Conversely, any prohibition of the possibility will involve formulas and equations. It has become somewhat fashionable to speak of mathematics as a "language." While it has many features of a language—mainly notation that looks like Egyptian hieroglyphics to nonmathematicians—there must be more than mere language going on. We still have to account for the amazing success of mathematics as a description of physical reality. The precision of so many theories of physical reality may hint at a deeper truth, that mathematics is a major structural foundation of our universe. Nothing expresses the presence of structure better than barriers. There are some things we cannot know and some things we cannot do, all as a result of the internal logical structure of this field. The limitation does not depend in any way on the wishes or fears of individual scientists, much less their cultural backgrounds.

For the sake of names I would invoke the entire universe of mathematics, both known and unknown, by the term "holos," Greek for "whole." The word "cosmos" has essentially the same meaning in Greek but, in this context, means the physical universe in which we live, move, and have our being. With such terminology one can make mystical-sounding statements such as "The cosmos rests upon the holos." I have only the vaguest idea of what this sentence might actually mean, but it advertises the aim of this book: to discover how physical reality depends on mathematical reality, and to examine how mathematical reality manifests itself—at least to exploring minds that are still capable of curiosity.

The physical world as described by physics has a somewhat eerie mathematical configuration, taken together. A stone thrown in a vacuum will execute a parabola with a precision great enough to rule out any other polynomial function as a possible path. Did Galileo and Newton lay this fantasy upon us because they were Italian or English? Because they were expressing a post-Renaissance yearning for perfection? Or were Galileo and Newton merely traveling the mental landscape of pure reason? Would Galileo recognize the thoughts of Newton? If not, why not? Both were explorers of the holos, and neither had much choice about what he would find—or not find.

The discovery of barriers to knowledge has been accelerating somewhat over the past two centuries, keeping pace with science itself. We have known since the late nineteenth century that we cannot square the circle, but in the 1920s we learned that there exist theorems that we shall never prove. In the 1940s, even as computers emerged from the smoke and dust of a world war, we discovered noncomputable functions and the unsolvability of the halting problem. This meant, among other things, that we would never be able write a program that debugs (finds all the errors in) other programs. In the 1970s we learned that even computable functions could be a problem. There are literally hundreds of well-known and important problems lurking behind every conceivable nook and cranny of our technological infrastructure. There are instances of these problems that cannot be solved by any computer (no matter how fast) before the universe comes to an end.

The foregoing barriers were all discovered by reasoning in pure mathematics. Although physicists may object, saying that I am laying claim to their territory, it can be suggested that the other barriers were discovered by reasoning in applied mathematics (or physics, if you wish). We have known since the mid-nineteenth century that we cannot build a perpetual motion machine, but learned, in the first ten years of the twentieth century, that we shall never travel through space any faster than the speed of light. In the 1920s we were dismayed to learn that we would never be able to measure precisely the simplest properties of tiny particles like the photon and the electron. In the 1980s, chaos descended on the field of dynamical systems with the discovery that the behavior of some not-so-tiny systems suffered from another form of unpredictability. We cannot predict the weather with anything like complete accuracy, nor will we ever be able to.

There are people among us who will brook no barriers, bridling at every limitation, as though it were their God-given right to be, well, God. Theology aside, there are others who, like me, will find marvels in these barriers to thought and action. A barrier gives shape, after all.

Like a landscape, science has spaces that we may freely roam, investigating phenomena and making progress. But we rebound from the walls that reason has discovered. "Adamantine" is too soft a word. Unlike cliffs or chasms, these cannot be penetrated by the intellectual equivalents of sledgehammer, bulldozer, or dynamite. They are insurmountable, impenetrable barriers. Reason brought these barriers to our attention, yet reason cannot penetrate them.

TWO OLD CHESTNUTS

For several centuries, it was thought that someone clever enough might just succeed in squaring the circle. Given a circle on a sheet of paper, construct a square with exactly the same area. The construction must be Euclidean, of course: the only instruments allowed are an unmarked ruler and a compass. Such means may seem unduly restrictive, yet with a ruler and a compass alone, we can construct a square equal in area to that of a triangle, to any square (of course), and to any pentagon, hexagon, or any regular polygon. The figures in this sequence come ever closer to resembling a circle. Who can distinguish between a circle and a thousand-sided regular polygon? Yet we cannot square the circle.

Although "squaring the circle" has become synonymous with all hopeless projects, the task was not known to be impossible until late in the nineteenth century, when the German mathematician Ferdinand Lindemann proved it so, once and for all. Before Lindemann's proof some of the finest mathematical minds in the world made attempts at the construction. All of them failed.

In the physical world similar barriers await us. For several centuries, some of the cleverest people around tried to construct perpetual motion machines. The rewards were, of course, enormous. Not only would the person who constructed the first perpetual motion machine become the most revered person in all of history (certain spiritual leaders aside), he or she also would stand a good chance of becoming the wealthiest person who ever lived. Mills and factories would all be run by such machines, and everyone who used them would, of course, pay the holder of the patent. With such huge rewards waiting, it is not surprising that more than a few had faked the results. But all of them failed.

Their devices were often wheels with rolling weights, with weights on pivots, running water, stationary or moving magnets, rods, belts, pulleys, and a variety of other attachments. Some of the devices made Rube Goldberg look like a rank amateur.

We may construct a machine that, once set in motion, will run for an arbitrarily long time. For example, we may build a succession of wheels that, once spun, continue to revolve for an hour, a day, or even a year.

What brings every wheel to a state of immobility, sooner or later, is friction. Even as our would-be perpetual wheel began to spin, we would hear a faint noise coming from it. It takes energy to make noise, the energy of friction, slowly bleeding away the wheel's momentum.

Of course, if we set a wheel revolving out in space, it will continue to revolve virtually forever, but that is not the precise meaning of perpetual motion. The device, besides continuing to run forever, must be capable of useful work. That is what all those unsuccessful inventors were aiming at.

Again, it was not until the mid-nineteenth century that we finally understood that the project was doomed. The newly emerging theory of thermodynamics dictated that the total energy of an isolated system was the sum of two components: potential energy (the energy of position) and kinetic energy (the energy of motion). In many of the proposed designs, potential energy was continually being converted into kinetic energy and back again, as though either process might, by some magic, increase the total energy of the system. Not so. Moreover, the new thermodynamics also stated that energy of an isolated system was a conserved quantity, neither created nor destroyed.

However, none of the proposed machines was an isolated system. The friction encountered by the devices also took part in the equation. The energy of motion included not only the moving parts of the proposed device but also the movements of molecules affected by it, those in the device and those in the surrounding environment. Although the total energy of the system + environment remained constant, more and more of it would be found in the agitation of molecules of metal, wood, and air. Slowly or quickly, the kinetic energy available for moving parts would decline.

More recently, we have encountered other barriers to what can be achieved by technology. As Einstein showed, neither we nor any signal we send can travel faster than the speed of light. Likewise, quantum mechanics, the most successful physical theory of all time, tells us that we cannot predict the detailed behavior of fundamental particles such as electrons and photons. We cannot even predict the weather, according to the theory of chaotic dynamical systems.

The latter barriers are not imposed by mathematics per se, but by well-established physical theories with a mathematical form that extends from physical axioms to physical theorems. Along the way, there are mathematical steps that invoke the grander edifice lurking in the background. We must, for example, use the Pythagorean theorem to get from Einstein's basic assumption (that light appears to travel at the same speed relative to all frames of reference) to the formulas of special relativity. Mathematics will have its way.

This book has accordingly been divided into two halves, the first concerning physical impossibilities, the second devoted to mathematical ones. This division reflects not only the cosmos and the holos but also the two major parts of science. Inductive science includes physics, chemistry, astronomy, and the other so-called experimental sciences. Deductive science includes pure mathematics, applied mathematics, the theory of computation, and related fields.

The process of induction involves inferring the rules governing the behavior of physical systems from particular instances. Knowledge accretes in a succession of increasingly general layers. For example, the properties of chemical compounds, once a confusing mélange of specific observations, became perfectly comprehensible from a unified point of view with the wide adoption of Dalton's theory of the atom.

The process of deduction, on the other hand, operates on abstract systems which, because they are perfectly defined in full generality, can be subjected directly to deductive operations. We already know everything we need to know about such systems in order to find out more. Here, knowledge builds outward from the axioms of a system toward all the specific theorems that are ultimately obtainable. And here, too, knowledge develops increasingly general layers. Seemingly unrelated things such as number systems, symmetries of crystals, and permutations of letters all came to be viewed as examples of groups, for example.

Scientific knowledge is about generality. One could say that the more general a successful theory is, the more "scientific" it has become. The most general fields of inductive science are heavily mathematical—in proportion to their generality, not surprisingly. Thus quantum physics and relativity theory are practically *all* mathematics—with an interpretive framework grounded in observations. In contrast, biology has relatively little mathematics in it, but a host of observational data that ecologists and biologists are still trying to make sense of. The two greatest discoveries of biology—the Darwin-Wallace theory of evolution and the Watson-Crick discovery of DNA—are among the most general (and mathematical) parts of biology. The Darwin-Wallace theory can be viewed as a simple deduction from axioms about reproduction, the tendency for populations to grow without limit, and survival. The Watson-Crick theory, on the other hand, describes a mathematical code that is the basic structure of every living creature. As pure mathematics, neither the deduction nor the code are of great interest, but their significances for science and humanity are immense.

All of this is not to say that at least one of these barriers may be found to be illusory, but we may expect most of them to be with us for a long, long time. Such a view, with its built-in back door, is only partially cowardly.

Theories are overturned, as Thomas Kuhn has rightly observed, from time to time. That much was known long before Kuhn, who introduced the term "paradigm shift" to signal more than a mere overturn. It was Kuhn who argued that scientific revolutions have their roots not so much in data, but in how we interpret them; theories are "social constructions." This is undoubtedly true in that social preferences surely favor one direction of research over another or one conceptual framework over a different one. Cultural bias may cause a scientist to miss a result, but someone else will probably find it.

But to claim that science is one big social construction amounts to throwing the baby out with the bathwater. In fact, Kuhn's interpretation of his main example, the Copernican revolution, is incorrect. The Ptolemaic notion of planetary motions was superseded by the Copernican model for a very good reason. The Ptolemaic theory, with its complicated system of epicycles, was simply wrong. Long before Copernicus, few astronomers were very happy with epicycles. Even in the post-Ptolemaic period, several Arab astronomers were certain it was incorrect. Anyone who thinks that a social construction lurks in the labors of Kepler (who put the theory on a solid footing with his discovery of elliptical orbits) should read of research driven purely by the observational data of Tycho Brahe, and learn of Kepler's frustration when he discovered that the one social construction he did attempt, the "mysterium cosmographicum," was a dismal failure. It did not fit the data.

Today, no one doubts the Copernican theory. It is the correct theory. The extremely slight changes induced by general relativity do not change the fact that the Earth goes around the Sun and not vice versa.

It has been said, "Never say never." Will the barriers described in this book never be penetrated, leaped over, or gotten around? At the end of each chapter, I examine what appear to be exceptions to the claim or discuss potential breakthroughs that would undermine it. For example, to do away with million-year space voyages that are limited by the speed of light, science fiction writers have invented the warp drive, a device that folds space. Two parts of the space-time continuum, previously far apart, are brought into proximity, and the starship *Enterprise* leaps halfway across the galaxy. Well, be careful! The limitations here are sometimes subtle. Even if we had such a spaceship now, it would not actually violate the cosmic speed limit. After all, it wouldn't travel *through* space, but *around* it, so to speak.

Limitations on what we can know or do, whether real or only apparent, have the salutary effect of driving the scientific and technological process forward. Based on the apparent acceleration of impossible findings, it seems safe to predict that science will increasingly become entangled with things unknowable and undoable. Perhaps all the barriers, taken together, will more sharply outline this place we find ourselves in.

MATH IN THE COSMOS

· 1 ·

The Energy Drain

Impossible Machines

> IT IS NOT POSSIBLE TO BUILD A MACHINE
> THAT RUNS FOREVER WITH NO SOURCE OF
> ENERGY, YET PRODUCES USABLE ENERGY.

LIKE SQUARING THE CIRCLE, the problem of making a machine that would run forever has probably absorbed more man-hours than the building of the Egyptian pyramids. The idea of the "perpetuum mobile" is nearly as old as the pyramids. It is so seductive that I run a certain risk even writing about the subject. I mean, why shouldn't it be possible to build a machine that runs forever? I'll even throw in the requirement that it produces a little usable energy. In my mind's eye I see a simple but wonderful contrivance that will (in my mind) run forever. It is nothing more than a wheel with beautifully curved spokes, each engraved with a track that carries a steel ball.

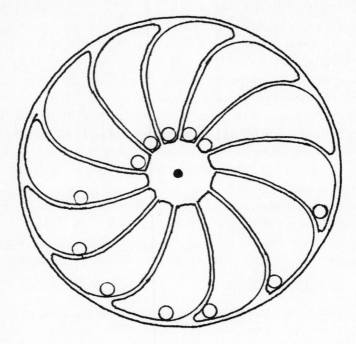

A perpetual wheel

Clearly, the balls on the right-hand side of the wheel shown here are going to weigh more heavily, in turning the wheel, than those on the left. Hence, with barely a nudge, the wheel will begin to rotate. As each new spoke comes into position, another steel ball rolls to the right. In fact, because of the continuing downward acceleration of the right half of the wheel, it will spin with increasing velocity. One might imagine that it will spin faster and faster, without limit, until it literally flies into pieces. But at some point during this potentially fatal acceleration, the centrifugal force on the balls, particularly those on the left side, will prevent them from rolling toward the hub, as they would normally do. At such a pass, the wheel will continue to spin, gradually slowing, owing to frictional forces. As soon as the wheel slows enough to allow the balls on the left half of the wheel to roll to the right once more, it will, of course, begin to speed up again.

The wheel will therefore seek and find, ultimately, the narrowest range of speeds wherein it will remain spinning like the governor of a steam engine. Unlike a governor, however, this wheel has no external source of power! It will spin at exactly this rate until the parts begin to wear out. Sooner or later, something will give way and the wheel will roll to a clattering halt.

Nevertheless, it has been a valid demonstration of perpetual motion, at least conceptually. We could construct the wheel and its parts of modern space-age alloys, with Teflon bearings and silicon lubricants. It might be guaranteed to run for a thousand years. Or a million. Put enough care into it, and the latest version of the wheel will run for a billion years. We are limited only by the lifetime of the universe, whatever that might turn out to be.

The time requirement of the problem—that the device operate perpetually—is clearly unrealistic. As I just hinted, there is rather strong evidence that the universe will one day cease to exist altogether, taking our wonderful machine with it. But in a purely theoretical sense, the machine is potentially capable. It would, if it could, spin forever.

Now, if it were possible to eliminate all friction from an ordinary wheel, it, too, would spin forever, thanks to a law first discovered by the great British natural scientist Isaac Newton. This motion, while perpetual within any practical meaning of the word, is not the sort of motion we have just been discussing, as it produces no new energy. I will therefore call it type one perpetual motion. Earth satellites are essentially type one perpetual motion machines.

In type two perpetual motion we expect the device not only to exhibit potentially eternal motion but also to produce energy while doing so. In my opinion, type two perpetual motion is the more exciting of the two kinds.

Can the wheel in my mind's eye do actual work? Think of my wonderful wheel again, now writ large.

It will be found inside a secret government building somewhere in the desert. The wheel will be forty stories tall and made of more than a million tons of steel, Kevlar, diamond, and other awesome materials. We'll take an elevator to the fifteenth floor of the building, near the hub. We'll stare in stupefaction from our observation window as, one by one, the great spokes swing by, each with a great steel ball rolling along it with a terrifying rumbling noise. The secret installation sounds like a cosmic bowling alley.

Now, just before the great wheel reaches that critical speed where centrifugal force begins to eat away noticeably at its acceleration, a giant dynamo engages gears with the wheel, and outside the building power lines surge with millions of kilowatts of power, all of it apparently free.

That's what I mean by energy. Even a type two machine that produces the tiniest excess of energy may nevertheless be scaled up to almost any size—or multiplied a thousandfold by mass production. Indeed, any macroscopic machine that runs "forever" under ordinary circumstances must be overcoming frictional forces and must be type two.

When scientists say that perpetual motion is impossible, they mean only that a machine that produces more energy than it consumes (typically none) is impossible. Type two machines are impossible. The reason for the impossibility lies with a fundamental tenet of modern physics, the law of conservation of energy. I will come back to this law later, subjecting it to a scrutiny it rarely receives. (Perhaps there is a flaw somewhere.)

In the meantime, we have been examining "machines" in the ordinary sense of the word, macroscopic systems of metal, plastic, even wood. But the microscopic world is inhabited by other systems: atoms in a state of apparent perpetual motion, electrons that whirl incessantly about nuclei, albeit in an unmanifest state, even at absolute zero. They are not producing any new energy, but their motion appears to be truly perpetual. After all, there is nothing to "wear out."

From this point of view, the closest we have yet come to a type one perpetual motion machine, here on Earth, is a doughnut-shaped ring of some hi-tech superconducting material in which electrons have been set circulating. Back in the 1980s, newspapers and science magazines frequently showed pictures of a superconducting ring floating above a magnet. How did it work?

The doughnut or torus is made from one of the latest metallo-ceramic hybrid materials that "superconducts" electrons at a temperature that is low, but not too low to be achievable in a laboratory. Superconduction works like ordinary conduction except that electrons move through the material without the production of heat (and subsequent loss of energy) that accompanies normal conduction.

Electrons in such a ring are set in motion by placing the ring in a strong magnetic field. The apparent levitation of the ring is not part of

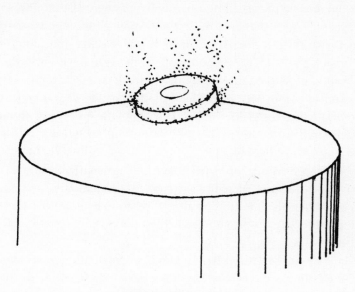

Superconducting ring

the perpetual motion of the electrons inside it. Perfectly balanced, the ring does not move nor even rotate. The levitation is merely a macroscopic signal that the electrons inside it are circulating, something that newspapers never explained and science magazines rarely.

The Lorentz effect is a well-known physical phenomenon in which an electron, moving at right angles to a magnetic field, experiences a sideways force in a direction that is simultaneously perpendicular both to its direction of travel and to the field. Continually bending in the same direction, it travels in a circle and will do so as long as the magnetic field is present. At the same time, electrons traveling in a circle generate a magnetic field of their own. In the case of the ring, merely moving it into the magnetic field sets the electrons in motion. The proof that they do not slow down, as they would if the ring were made of an ordinary conductor, is the continuing opposing force that the circulating electrons set up. The magnetic "north" of the magnet is opposed by a balancing "north" generated by the electrons in the superconducting torus. It floats eerily in the magnet's field.

Left to themselves under ideal conditions, the electrons might be coaxed into circulating for arbitrarily long periods of time. This is perpetual motion of sorts, but it involves a microscopic motion within a macroscopic object that does not actually move, but remains suspended. In any event, the electrons have no frictional or other opposing forces to overcome. If they did, they would quickly grind to a halt, and the magic ring would fall with a rattle to the magnet beneath it. It follows that the circulating electrons are incapable of producing any new energy, and the superconducting torus exhibits, at best, type one perpetual motion. In the traditional search for perpetual motion, only machines with macroscopic motion need apply.

It fascinates many people, particularly mathematicians, how various physical theories interact in ways that their framers might not have foreseen. When they do, the result is rarely a contradiction. For example, suppose some bright young scientist developed a gravity-blocking material.

Such a device was used by the famous British science fiction writer H. G. Wells in his novel *The First Men in the Moon*. Two adventurous souls sit inside a special sphere equipped with food, air, and the other necessities of life. Two sets of blinds made of "cavorite" (after one of the adventurers, a scientist) are used to propel the sphere toward the Moon. By simply drawing the Earthside blind, gravity is blocked and the sphere floats upward, attracted by the Moon's gravity. Close to the Moon, the hardy explorers simply close the Moonside blind and open the Earthside one to slow the sphere to a gentle landing.

Is a material like cavorite possible? If it were, I could instantly invent a perpetual motion machine. I would simply turn a bicycle upside down and place the cavorite sheet on the ground below the front half of the front wheel. The portion of the wheel screened from gravity by the cavorite sheet would be weightless, and the heavier half would begin to turn Earthward. The wheel would therefore spin, faster and faster, until it reached its mechanical limits. Moreover, it would produce energy and be a true type two perpetual motion machine. The resulting contraption appears to contradict the laws of conservation of energy—until we realize that it resulted from the assumption that cavorite is possible. Cavorite or any material or device with the same property must therefore also be impossible. We have thus discovered a strange link between perpetual motion and gravity!

But is perpetual motion really impossible? If so, does this not represent an intolerable limitation on our freedom?

TYROS AND TRICKSTERS

The wheel on page 12 was actually dreamed up in the 1640s by Edward Somerset, the second marquis of Worcester. He had a device built, a monstrous wheel with forty 50-pound balls (weighing in total 1 ton) that rolled along curved spokes of wood. In 1648 he demonstrated his machine to King Charles I and members of the royal court in the Tower of London. Somerset was wealthy, but not exactly a dilettante. He contributed to the early development of the steam engine by designing a two-chambered engine that could pump water using the action of steam to create a partial vacuum.

He describes the occasion of the royal inspection of his wondrous machine as follows: "The wheel was fourteen foot over, and forty weights of fifty pounds apiece. Sir William Balfour, then Lieutenant of the Tower, can justify it with several others. They all saw that, no sooner these great weights passed the diameter-line of the lower side, but they hung a foot further from the centre, nor no sooner passed the diameter-line of the upper side, but they hung a foot nearer. Be pleased to judge the consequence."

Whatever the "consequence" was, it seems doubtful that the king or any of his courtiers saw perpetual motion that day. A wheel of such size and weight, once set in motion, might turn for a considerable time before stopping. And if, long after the royal party had departed, the wheel slowly came to a halt, the marquis might attribute the phenomenon to poor construction. All evidence points to Somerset's sincere belief that he had, indeed, achieved perpetual motion.

Was there a link between the development of the steam engine and the craze over perpetual motion? Did people once think that if only they were clever enough, they could dispense with the messy boiler and the need for fire? Probably not. The dream of perpetual motion goes back much earlier than the seventeenth century.

An intriguing passage in a seventeen-hundred-year-old Sanskrit book on astronomy, the *Siddhanta Ciromani,* describes a self-turning wheel. Its outer rim or tread was to be drilled with equally spaced holes that (probably) pointed not toward the axis, but a little ahead of it, so to speak. Each hole was half filled with mercury and then sealed. The passage claims that if the wheel was properly supported, it would turn forever.

In spite of the similarity in principle behind the Indian device and the wheel I have been describing, it is doubtful that any of the European Renaissance inventors were familiar with the Sanskrit book. Mark Antony Zimara, an Italian physician and astrologer, published a treatise on diseases of the body in about 1520. As an incidental enclosure, he included a description of a perpetual motion machine (something that few medical texts could get away with these days). The machine consisted of a large fan, which, when it turned, worked several levers that, in turn, operated three giant bellows. The bellows, in case you hadn't guessed by now, blew directly on the fan to keep it going. No record exists of anyone attempting to build Zimara's machine, and Zimara himself may have had lingering doubts, as this translation from the Latin hints: "This, perchance, is not absurd, but is the starting point for investigating and discovering that sublime thing, perpetual motion, a starting point which I have not read of anywhere, neither do I know of anyone who has worked it out."

Most of the early designs for perpetual motion machines in the Renaissance involved feedback of one kind or another: the motion of component A serves to keep component B moving, which, in turn, powers A.

In 1618, this principle was actively employed by Robert Fludd, an English physician and philosopher. He designed a closed-cycle grinding mill for which the motive power came from an endlessly circulating flow of water over a mill wheel. The wheel not only ground the grain but also powered an Archimedean screw, as in the illustration.

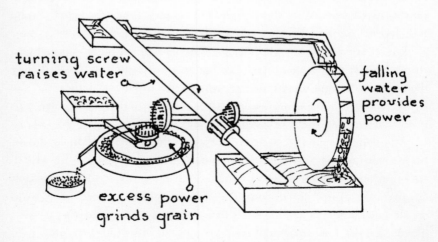

Fludd's closed-cycle grinding mill

Once set rotating, the screw brought the water from a discharge basin up to a reservoir, where it could begin its trip to the wheel all over again. Clearly, Fludd thought that somewhere in the cycle of the mill's action, the system would gain energy. Would the energy of the falling water be greater than the energy it took to transport the water to the top of the millrace? It is generally easier to propose a grandiose scheme than to carry it off. Fludd made no attempt to build his perpetual mill. If it had worked, Fludd's machine would have been a good example of a type two perpetual motion machine.

Many clergy of seventeenth-century England were also amateur scientists and inventors. Bishop John Wilkins of Chester designed a scheme very similar to Fludd's but went to the trouble of actually building it. He had the integrity to make his failure public: "When I first thought of this invention, I could scarce forbear, with Archimedes, to cry out 'Eureka! Eureka!' it seeming so infallible a way for the effecting of the perpetual motion that nothing could be so much as probably objected against it; but, upon trial and experience, I find it altogether insufficient for any such purpose, and that for two reasons: (1) The water that ascends will not make any considerable stream in the fall. (2) The stream, though multiplied, will not be of force enough to turn about the screw."

One of the most famous perpetual motion machines was constructed by a Polish German engineer, Johann Ernst Elias Bessler, who styled himself as "Orffyreus." In the 1710s, all Europe was abuzz with news of Orffyreus' amazing perpetual wheel. It was housed in a room of a castle that belonged to the landgrave of Hesse-Cassel at Wissenstein. The landgrave allowed Orffyreus to set the wheel in motion, then locked the room in which it was kept, with the landgrave's own official seal on the lock. At the end of two months, the story went, the room was opened, and there was Orffyreus' wheel, still turning at a good clip.

Many distinguished philosophers, engineers, and scientists came to view the amazing machine, and all went away shaking their heads in wonder. Writing to the French philosopher Jean Desaguliers, Baron Fisher described what he saw: "The wheel turns with astonishing rapidity . . . twenty turns a minute. This I noted several times by my watch, and I always found the same regularity. An attempt to stop it suddenly would raise a man from the ground. Having stopped it in this manner it remained stationary (and here is the greatest proof of perpetual motion). I commenced the movement very gently to see if it would of itself regain its former rapidity, which I doubted; but to my great astonishment I observed that the rapidity of the wheel augmented little by little until it made two turns, and then it regained its former speed."

Orffyreus' wheel was 12 feet in diameter, 14 inches wide, and had a rather large axle, about 8 inches thick. The entire wheel was covered with a waxed cloth to conceal the inner workings from those who might seek to steal the great Orffyreus' idea. After receiving a handsome gift from the landgrave, Orffyreus showed his patron the interior of the machine. He had extracted a solemn oath from the landgrave that he would never divulge the marvelous mechanism.

Perhaps what he saw roused the landgrave's suspicions. And perhaps enough skeptics of the wheel's operation remained to prompt him to launch an investigation. He hired the Dutch philosopher and engineer Professor s'Gravesande of Leyden to investigate. We have the professor's findings in a letter to Sir Isaac Newton. He found the wheel "covered over with canvas, to prevent the inside from being seen. Through the centre of this wheel runs an axis of about six inches diameter, terminated at both ends by iron axes of about three-quarters of an inch diameter upon which the machine turns. I have examined these axes, and am firmly persuaded that nothing from without the wheel in the least contributes to its motion. When I turned it but gently, it always stood still as soon as I took away my hand; but when I gave it any tolerable degree of velocity, I was always obliged to stop it again by force; for when I let it go, it acquired in two or three turns its greatest velocity, after which it revolved for twenty-five or twenty-six times in a minute."

Orffyreus, who was apparently not informed of the inspection, flew into a rage at the news, went to the castle, and smashed his marvelous engine beyond all repair. He never built another wheel, as far as I know, and no one ever explained how the wheel was able to turn so long without an outside source of energy. Suffice it to say that Orffyreus was trained as a clockmaker. It might be suspected that the extraordinarily thick axle held the secret. In any event, the need to conceal a secret source of energy would be covered by the same story as the need to protect a genuine idea. After all, a working mechanism would be worth, ultimately, all the wealth of the world, and no attempt to conceal its secret, however elaborate, could be considered unreasonable.

In the nineteenth century the search for perpetual motion intensified, amounting almost to a craze. At the same time, the skeptical eye had been honed to a point that would have made an Orffyreus impossible. The scene shifts to America, where, in 1812, a man named Charles Redheffer appeared in Philadelphia with a curious machine that, he claimed, would never stop. The public, eager for wonders, flocked to see the machine. Bets, some of them quite large, were made over the proof or disproof that Redheffer's machine actually worked as claimed.

The intrepid inventor applied to the Pennsylvania legislature for what today we would call a research grant. To inspect Redheffer's machine to determine the feasibility of perpetual motion, the legislature duly dispatched eight commissioners to examine the device. When they arrived at the building that housed the machine, they found that the room containing the machine was locked and Redheffer was nowhere in sight.

Through a barred window they saw a turntable mounted on a spindle shaft and apparently powered by two miniature trucks that rested, motionless, on two inclined planes, respectively. Each truck contained two small weights that, it was said, provided the actual power owing to their attraction by gravity. Although the trucks did not move, they were attached by levers to the spindle. The levers thus transmitted the force of gravity to the spindle, turning it—or so Redheffer claimed.

It is amazing that so many people were taken in by this machine. Since the little cars did not move relative to the large supporting wheel, nor the wheeled inclined planes, it would be utterly mysterious to a layperson untutored in mechanics how the thing worked. (It would, of course, be even more mysterious to a knowledgeable engineer.) Perhaps the gimcrack nature of the machine, coupled with the hypnotic effect of the slowly turning wheel, succeeded in convincing not only the lay visitor but the professional engineer as well. I mean, there it was, *turning*.

The illusion was reinforced by Redheffer or whomever he hired to operate the machine. Removing the little weights brought the device to an immediate halt.

The inspection by the commission of engineers employed by the legislature included a gentleman named Nathan Sellers, who brought his son along to see the machine. In the course of the inspection through the barred window, the son tugged on his father's coat. "See, Papa, the gearing looks wrong." The lad was referring to the birdcage gear that was supposed to be driven by the great wheel. The boy pointed out to his professional father, as well as to the other gentlemen present, that the wear on the wooden teeth of the birdcage gear was on the wrong side.

In other words, the gear was driving the great wheel, not the other way around.

Saying nothing to Redheffer, Sellers consulted later with Isaiah Lukens, an engineer and skilled mechanician with the Franklin Institute in Philadelphia. Sellers described the machine well enough for Lukens to make an almost exact copy. This would function as a lure to draw Redheffer out and so expose the fraud.

Although Lukens could not know what ultimate source of power Redheffer drew upon to keep his machine in motion, he devised one that was no doubt equally subtle. Under the handsome wood paneling below the great wheel, he installed a small spring motor that could be wound up by turning one of the wooden knobs on top of the ornamental framework that surrounded the machine. Thus a custodian could approach the exhibit as the machine was running down, draw out a rag and pretend to dust the machine, in reality turning the secret knob a few times as he pretended to polish it.

A demonstration of the duplicate machine was arranged by Sellers and Lukens, and they invited Redheffer to attend. Lukens had even arranged that the duplicate machine would also halt when the weights were removed. The clockwork motor concealed in the machine did not drive the birdcage gear, but the main axle itself, transmitting its force through friction that was barely sufficient with the weights in their cars. But with the weights removed, the friction dropped below the critical value, and the turning of the motor had no effect on the big wheel.

Redheffer could not conceal his amazement at the device shown to him by these sober, respectable citizens. Privately he offered Sellers a great deal of money if only he would reveal the principle by which the machine operated. Sellers may have replied, "Why, the principle is the same as that employed by your own good self: chicanery!"

Exposed, Redheffer decided to move his exhibit to New York. He had fleeced thousands of Philadelphians out of their one-dollar admissions (a good deal of money in those days) and could depend on the poor communications of the day to delay news of the scam.

Unfortunately for Redheffer, New York was not only the home of a much larger flock of potential believers but also the lair of one of America's foremost engineers, Robert Fulton. Although he refused at first to attend the exhibit of what he considered to be an obvious fraud, friends prevailed upon him, and Fulton finally gave in. As soon as he entered the room to inspect the machine, he cried, "Why, this is a crank motion!" He meant that the unevenness of the turning sound from the device indicated that someone, somewhere, was turning a crank.

He denounced Redheffer, who happened to be present, on the spot. Redheffer blustered and became angry. Fulton said he would repay Redheffer in full for all damages and proceeded to attack two supports that were presumably meant to steady the machine against one wall. Inside one of the supports, Fulton discovered a catgut drive belt. Like bloodhounds on the scent, Fulton and his cronies sniffed at a hole in the wall through which the drive belt passed. They forced their way through a series of rooms until they came, according to one account, upon "an old man with a long beard who displayed all the signs of having been imprisoned in the room for a long, long time. The man had no notion of what was happening and sat there on a stool gnawing on a crust with one hand and turning a crank with the other."

Although surely a romantic exaggeration, the discovery of an accomplice put the crowd into a rage. They destroyed Redheffer's marvelous machine, and the great inventor had to flee for his life.

Perpetual motion machines continued to be the stock in trade of tricksters and con artists throughout the nineteenth century. In the 1850s, an engineer from Connecticut, C. P. Willis, made an elegant horizontal toothed wheel of brass that was enclosed, with the rest of the mechanism, in a glass case, where the curious could ponder the endless motion of the wheel. As the brass wheel spun, it engaged a gear that turned a flywheel that communicated with the first wheel by a system of pulleys. It was the old dream of A drives B while B drives A.

Willis charged people to inspect his machine, first in New Haven and later in New York. An alert patent attorney, however, noticed a strut near the flywheel that had no apparent function. He discovered that Willis had arranged a supply of compressed air, fed into the strut, to blow on the flywheel and keep it in motion.

It would be fascinating to view a vast museum exhibit of all the perpetual motion machines ever built or even imagined. There would be hundreds, if not thousands. There would be not only overbalancing wheels and feedback water mills but also sponge machines employing capillary action, machines driven by magnets, and numerous other devices. They could all be started at once, and the great exhibit hall would be filled with a kind of mechanical sigh of despair as collectively and one by one the machines ran down or their fraudulent energy sources were cut off.

PROOFS OF IMPOSSIBILITY

Newton's third law states that a body at rest or in a state of uniform motion will continue in the state unless influenced by some external force. In other words, all motion is perpetual in the absence of external forces. Paradoxically, type one perpetual motion is built into the very foundations of classical physics.

As for type two, or true perpetual motion, one may analyze each device in turn and prove that it could not possibly be producing more energy than it consumes. Such proofs, because of the complexity of some devices, may involve considerable labor. In other cases, as below, the analysis is relatively straightforward. I will present an analysis of a device first described by the Dutch philosopher/mathematician Simon Stevinus at about the turn of the sixteenth century.

As far as I can tell from available sources, Stevinus may have "invented" the machine that appears in the figure to demonstrate the impossibility of perpetual motion (in this one case), or he may have come upon the design in sources that are no longer available. I prefer to think the latter.

The machine shown here consists of fourteen "rollers," essentially cylinders that are linked by a flexible chain connecting their axles. Four

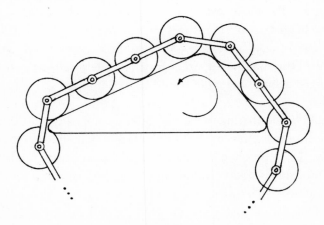

The two-track roller machine

rollers occupy the long ramp on the left of the figure, while two rollers rest upon the shorter ramp on the right. It is easy to imagine that the four rollers on the left exert a greater force on the endless chain than do the ones on the right. After all, there are four of them. As Stevinus observed, the eight rollers (not shown in the figure) that hang under the ramps make no contribution to the resulting perpetual motion because they exert an equal force on both the right and left portions of the chain. My analysis will be a simplified and more modern version of the one given by Stevinus.

Let us call the long ramp A, the short one B. We do not know the exact lengths of these ramps, but if we designate them by the symbols a and b, respectively, then $a = 2b$, since the long ramp accommodates twice as many rollers as the short one.

To analyze this particular machine, we will search for the motive power that would result from an imbalance of forces among the rollers. Such an imbalance would surely manifest at the portion of chain joining the four rollers on the long ramp to the two rollers on the short one. At that point, the force due to gravity manifested along the long ramp will presumably be greater than the force along the short one.

We use the trigonometric cosine function to express the action of gravity along the angle made by either ramp with the vertical. In other words, although the force of gravity acts downward on a roller, the ramp

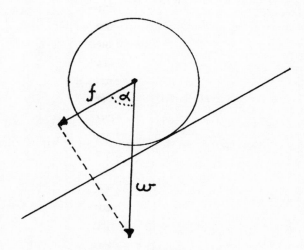

Diagram of forces on a roller

prevents it from moving in the downward direction. The force must act along the ramp, that being the only direction in which a roller can move.

In this right-angled triangle the vertical arrow represents the downward force on a roller due to gravity. If each roller weighs w grams (or pounds), the downward force is w. The sloping arrow, on the other hand, represents the component of gravity acting on the roller in the down-ramp direction. We'll call this force f. If α is the angle made by the ramp with the vertical, then the cosine of α, written $\cos(\alpha)$, is simply the ratio of the sides f/w. In other words, $\cos(\alpha) = f/w$, so that f, the downramp force, equals $w\cos(\alpha)$.

If α is the angle made by the long ramp with the vertical, we'll suppose the short ramp makes the angle β. The downramp force to the left must result from the weight of four rollers, as resolved by the cosine operator:

$$4w\cos(\alpha).$$

Similarly, the downramp force in the other direction, with just two rollers, must be:

$$2w\cos(\beta).$$

We can now eliminate the cosine function by recognizing that it represents the ratio of two distances. In this case, $\cos(\alpha) = c/a$, while $\cos(\beta) = c/b$. The two forces may now be written as follows:

$$\text{force along shallow ramp} = 4w\cos(\alpha)$$
$$= 4wc/a,$$

while

$$\text{force along steep ramp} = 2w\cos(\beta)$$
$$= 2wc/b.$$

As we observed earlier, the distances a and b have the relationship

$$a = 2b.$$

This leaves us with the

$$\text{force along shallow ramp} = 4wc/a$$
$$= 4wc/2b$$
$$= 2wc/b,$$

which just happens to be the force along the steeper ramp. Obviously, the rollers will not begin, of themselves, to move. Nor will they develop any additional energy as a result of moving. If friction is reduced to a minimum, the rollers may continue to circulate up one ramp and down the other for a while, but eventually they will slow down and stop.

A similar analysis could be made of the Somerset machine, except that the geometry leads to more complicated expressions. Suffice it to say that for each position in which the wheel may be overbalanced in one direction, there is another position in which it overbalances by the same amount in the opposite direction.

If a mere glance at the first position was enough to convince Edward Somerset, as well as thousands of people after him, that the wheel would turn forever, a glance at the second position might have given him pause. If popular illustrations of the wheel had included both positions, the machine might not have inspired such perpetual emotion.

PHYSICS HAS ITS SAY

There are general physical laws that imply that type two perpetual motion is impossible. The laws may even be cast in mathematical form, as we will presently see. Thus, in analyzing the various proposals for perpetual motion machines, we may always say, "There's no need to analyze these things in detail. There are physical laws that say such machines can never work."

On the other hand, we may also analyze each machine mathematically, employing the simplest physical concepts such as leverage and force, concepts that have nothing to do with the grand laws we are about to explore. Applying these concepts and incorporating them into a mathematical analysis of each device always lead to the same conclusion: it cannot work as advertised.

This is a very strange phenomenon when you think about it. It implies a consistency between our applied mathematical analyses and a general fiat about conservation of energy, a consistency that cannot be explained within any framework of knowledge currently available to us. If we had no notion whatever that all physical systems were constrained by the law of conservation of energy, we would still be reaching conclusions of impossibility for these machines based purely on applied mathematical reasoning.

CONSERVATION OF ENERGY

The historical beginnings of the law of conservation of energy reveal an ironic twist. The law was inspired in part by the belief of early scientists, beginning with Galileo Galilei of Italy, that perpetual motion was impossible. Based as it was on the continuing failure of such devices to work, we cannot say that the belief was exactly arbitrary. However, it led to some astonishing progress from the seventeenth to the nineteenth centuries.

Galileo, for example, analyzed the motion of bodies rolling down inclined planes. He concluded that a ball rolling down a straight ramp from top to bottom must reach the same velocity (friction aside) at the point where the ramp met the tabletop as it would if dropped directly from the release point onto the table. If this were not so, he reasoned, type two perpetual motion would be possible. Suppose the ball came off the ramp at a greater speed than it would develop by merely falling vertically. In such a case it could be deflected upward by a suitably arranged elastic barrier at the bottom of the ramp. The ball would bounce upward, rising higher than the point from which it started. If it landed on a second ramp, it could roll back to its starting point on the first ramp with even greater energy than on the first occasion.

On the other hand, if the ball came off the ramp with a slower speed than it would have if dropped (friction aside), the procedure outlined above could be reversed. If the ball were dropped from the same height as the starting point on the ramp, directly onto the same barrier, now arranged to deflect the ball directly up the ramp, it would travel beyond the release point. In this case it could be arranged for the ball to fall through a second hole, striking the tabletop with even greater force than it did on the first occasion and so on, ad infinitum.

An able experimenter, Galileo tested his hypothesis and found it to be correct.

Two hundred years later, Sadi Carnot, a brilliant young French scientist, devised a conceptual scheme that gave direct insights into the relationship between mechanical work and another form of energy: heat. In 1824 he imagined a peculiar apparatus consisting of a gas-filled cylinder with a piston that fitted the cylinder so perfectly that no gas could escape. Two heat reservoirs, one at a high temperature, the other at a low one, served to add heat energy to the cylinder or to remove it when the cylinder was placed over them.

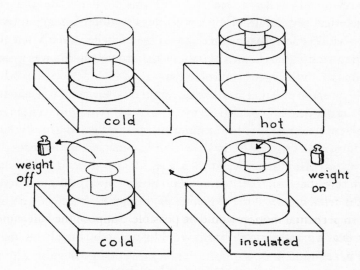

Carnot's "gedanken" experiment

Placed over the cold reservoir, the piston would take up a particular position within the cylinder. Placed over the hot reservoir, the cylinder would gain heat and the temperature of the gas would rise, accompanied by a rise in pressure. The piston would therefore rise in the cylinder to a new, higher position.

After heating his conceptual cylinder, Carnot moved it to an insulating pad, where the piston's position remained unchanged. Carnot then placed a weight on the piston, causing it to descend into the cylinder back to its original position. Next he moved the cylinder to the cold reservoir, simultaneously removing the weight. The position of the piston did not change in the end. The tendency of the piston to rise, thanks to the removal of the weight, was exactly counterbalanced by a drop in pressure within the cylinder, thanks to the removal of heat.

Heat had therefore been transferred from the hot reservoir to the cold one, the exact amount being equivalent to the work done by gravity acting on the weight and causing the piston to compress the gas. Carnot concluded that there was a direct and simple relationship between changes in heat energy in a cylinder and the amount of work done on it. Today we call this conceptual process the Carnot cycle, fundamental for understanding everything from automobile engines to refrigerators. Heat and mechanical energy are simply two forms of energy, and one could be converted into the other. Except for loss through leakage of heat or friction acting on mechanical motion, energy was conserved. The equations governing the relationship of mechanical and heat energy remain fundamental to thermodynamics today.

It is germane to our central theme that the experiments of Galileo and Carnot had a purely logical (i.e., mathematical) structure. Whether the experimental system involved rolling spheres or heated gases, a logical deduction could be made from the assumption that energy was conserved. Here is how the system ought to behave. Voilà! It does. Interestingly enough, the conclusion that conservation of energy is real is not actually a logical deduction from the experimental results. Only if the experiment had turned out negatively could either Galileo or Carnot have said, "Alas, energy does not appear to be conserved!" Instead, we can only say that the experimental outcomes "support" the idea that energy is conserved. That is the real difference between inductive and deductive science.

In the next chapter we will see an experiment in which no actual test is necessary, the famed "gedanken" experiments of Einstein and associates.

A mere twenty-three years after Carnot published his analysis of heat engines, the German physicist Hermann von Helmholtz presented the first formal statement of the law of conservation of energy. He began his address to the Physical Society of Berlin by declaring that perpetual motion was axiomatically impossible. The general law he proposed stated that energy could neither be created nor destroyed, but only changed from one form to another.

These early foundations of the law of conservation of energy might therefore be cited as a form of circular reasoning: perpetual motion is impossible because conservation of energy is true; conservation of energy is true because perpetual motion is impossible. On the other hand, if perpetual motion really was impossible (and empirical experience strongly suggested this to be the case), then additional weight was added to the newly minted law.

One could say that the search for perpetual motion, by focusing the attention of scientists on the intimate energy transactions within various machines, caused them to consider the same transactions within the wider scope of all physical systems. If the search for perpetual motion was not exactly the mother of conservation of energy, it was certainly the midwife. In any event, we have come to the final act of this drama: What, exactly, does conservation of energy mean?

When we talk about the conservation of energy, we must understand what we mean by "energy." Everyone understands that a fast-moving object such as a speeding automobile has a lot of energy. Just try to stop one. All motion involves the form of energy we call "kinetic." Even very small objects such as atoms carry kinetic energy when they move.

Another form of energy is heat. In most physical bodies it is stored in the form of kinetic energy, namely the motions of the many atoms that compose them. Heat may also take the form of radiation, being received or emitted by a body in the infrared portion of the electromagnetic spectrum. Machines with moving parts constantly generate heat from friction. This heat is normally conducted or radiated away.

There are other forms of energy as well. Potential energy is the energy of position. It depends on an external force field such as gravity, electricity, magnetism, or what have you, to be expressed. A car on a 100-foot cliff may not be moving, but relative to the ground below it, the car carries the potential to develop a lot of kinetic energy. Push it off the cliff and you will witness the effect of the Earth's gravitational field on the vehicle. Its potential energy will be converted into kinetic energy that is fully equivalent to the energy of the car when it speeds down the highway. All bodies, from planets to atoms, have potential energy of one kind or another.

Energy, as it turns out, inheres not only in the kinetic or potential behavior of physical bodies but in their very substance. Mass, according to Albert Einstein's celebrated formula, is equivalent to energy:

$$E = mc^2.$$

The "rest energy" of a physical body equals its mass multiplied by the speed of light squared. This speed, expressed in meters per second, is very large even when not squared:

$$c = 299,792,458 \text{ meters per second.}$$

This may be written approximately as 3×10^8 meters per second. A kilogram of mass of any kind (steel, pigeon feathers, or garden soil) would therefore contain 3×10^{16} joules of energy. A joule of energy is well within our ability to experience directly. For example, it takes about 10 joules of energy to raise a 1-kilogram (about 2.2-pound) rock 1 meter (a bit over 3 feet) above the floor. If the same rock were converted entirely into energy, it would produce approximately 3×10^{16} joules. If this energy were released all at once in the form of an explosion, it would rival a very large atomic bomb in its effect. (Although atomic bombs use much larger masses, only about 0.1 percent of their masses are converted directly into energy.) A more striking example would involve an ordinary penny, which, if it could be converted entirely into energy, would supply the power needs of the average house for a lifetime.

We therefore have three kinds of energy. The law of conservation of energy states that in every "isolated" physical system, energy may be transformed from one kind to another, but never created nor destroyed. A proposed type two perpetual motion device, if not allowed any energy input from the outside world, would qualify as an isolated system but would not be exempt from this law.

Put mathematically, the law of conservation of energy states that an isolated physical system must obey the following equation:

$$mc^2 + K + P + H = E \text{ (a constant)}.$$

Here, mc^2 is the rest energy of the system. Assuming that no nuclear processes are involved in a perpetual motion device, this quantity will remain constant, and I can move it to the right-hand side, absorbing it in a new constant, E'. The letters K, P, and H represent kinetic, potential, and heat energy, respectively, within the system. A simplified equation results:

$$K + P + H = E' \text{ (another constant)}.$$

In Somerset's wheel, we learned of two positions from which it would turn spontaneously, albeit in opposite directions. In either position, the potential energy was high and the kinetic energy was zero. As the wheel began to turn from the first position, its value of K would rise, exactly compensated by a drop in P. As it continued to turn, it would approach the second position, at which point P would again begin to rise and K to drop, again by an amount that compensated exactly for the drop in K.

Did I say "exactly"? If there were no friction at all, either from the air or from the axle, this would be true. The wheel would approach the second position, pause, then begin to turn in the other direction, replaying the previous scenario in reverse. In fact the wheel would develop perpetual motion, but not exactly what Somerset envisioned. It would enter a regime of perpetual (type one) motion.

Since the wheel must encounter friction as it turns, it would not quite reach the second position. Nor, in turning back, would it ever again reach the first. Instead, it would rock back and forth in gradually diminishing arcs until it finally came to rest in what physicists call a position of equilibrium. Full stop. The energy robbed from the wheel by friction would be converted into heat $H,$ some of it retained in the wheel, the rest radiated away as electromagnetic energy. A diagram of the wheel in this final position would create very little excitement in potential inventors.

IS THERE A WAY AROUND IT?

The final question leaves us with an unsolved mystery, one that mathematics would seem unable to address because it involves a very deep structural property of the universe in which we find ourselves. How do we know that the law of conservation of energy is valid? As with other physical "laws," we do not make deductions, but inferences.

The only loophole I can think of involves the exchange of energy and information. Maxwell's demon, a conceptual imp named after the Scottish physicist James Clerk Maxwell, plays a prank on us by sitting astride a tiny doorway between two gas-filled vessels, A and B. The demon is small enough to observe directly the motions of individual gas molecules in either vessel. When he sees a molecule approach the doorway from vessel A, he opens the door and allows it to pass into vessel B. But when he sees a molecule approach the doorway from vessel B, he keeps the door closed. Thus, over time, pressure builds in vessel B until it contains virtually all the gas molecules.

A pipe that connects the two vessels passes through a turbine and, when a stopcock is opened, the turbine runs for a while until the pressures in the two vessels are once again equal. Again the demon sets to work, and again the turbine is run. Is this not—conceptually, at least— an example of type two perpetual motion?

The answer lies in quantum mechanics. As we will see in chapter 3, the kind of knowledge to which the demon is privy is impossible to obtain. He may know the position of a particle only to the extent that he is ignorant of its speed. He will not know when to open the door, in effect.

And even in quantum mechanics, energy is conserved, being quantized—that is, occurring in discrete packets called quanta. Each atom contains a certain amount of energy in the levels of excitation of its electrons. Should a single electron drop from a higher to a lower energy level, the atom loses energy. But the energy lost is emitted by the atom in the form of a quantum of energy that travels in the form of a wave until it is absorbed by some other atom. Quanta do not disappear, nor do they appear out of nowhere. Quantum mechanics is the most successful physical theory ever developed.

As one of the most general physical laws we have, conservation of energy pervades both the classical and the quantum mechanical branches of physics. In the classical domain, which for our purposes consists of Newtonian physics and relativity theory, no experiment ever performed has found a violation of this law. In fact, the mathematical accounting that goes into every physical calculation has never revealed a violation of either of the equations involving K, P, and H. Energy may be converted from one form to another, but it is never lost. Energy radiated away from one system will eventually arrive at another, to put it crudely. It is never lost in transit, nor does it ever appear out of nothing.

If someone, someday, designs and builds an actual type two perpetual motion machine, that person will not only become wealthy enough to make Bill Gates look poor, but also he—or she—would be entitled, I should think, to a Nobel Prize in physics.

· 2 ·

The Cosmic Limit

Unreachable Speeds

> IT IS NOT POSSIBLE FOR MATTER OR ENERGY TO TRAVEL FASTER THAN THE SPEED OF LIGHT IN A VACUUM.

EVERYONE KNOWS WHAT LIGHT IS, and yet no one knows. It shines brilliantly or dully on our world from many sources. The Sun, like most stars, emits light. The Moon and the planets reflect it. Fire made light for the ancients, and electricity makes it for us. And everybody has experienced these things, more or less.

Physicists also know what light is, more or less. It travels at the incredible speed of 299,792.458 kilometers per second. Moreover, it has a dual nature, consisting of particles or waves. Which aspect of light you see depends on the dimensions of your vision. In the macroscopic world of ordinary experience, where objects all have dimensions of millimeters or

35

more, all the phenomena we see can be explained by the wave nature of light. But in the microscopic world of the atom, where sizes and distances are measured in nanometers, light takes on a grainy, particulate quality. Light is "quantized," in a word, and its behavior in such tiny dimensions is truly bizarre, hiding many secrets behind a quantum curtain—to be drawn aside in the next chapter.

Here we focus on its wave nature and two discoveries. The first discovery, made by the Danish astronomer Olaus Roemer in the eighteenth century, was that light traveled at a finite speed. The second discovery, made by the German Swiss scientist Albert Einstein in the twentieth century, was that nothing could travel faster.

Both discoveries overturned earlier ideas about light in particular and motion in general. Aristotle had held that light was transmitted instantaneously from a luminous body, the cause of light, to its perception in the eye, the effect of light. He argued with those who believed that light had motion: "Empedocles . . . was wrong in speaking of light 'traveling' or being at a given moment between the Earth and its envelope, its movement being unobservable by us . . . if the distance traversed were short, the movement might have been unobservable, but where the distance is from extreme east to extreme west, the draught upon our powers of belief is too great."

Plato thought that light traveled, but in the opposite direction, so to speak, being an "influence" that left the eye and embraced the object seen. The Platonists who followed in Plato's footsteps developed a sort of hybrid theory in which light traveling from an object toward the eye was met and assisted by an emanation from the eye.

The earliest Greek theory of light, first advanced by the Pythagorean school, has a strangely modern ring. Light traveled with a finite speed, but consisted of particles too fine to see. This amounted to a corpuscular theory of sorts, which the English scientist Newton would revive some twenty-three centuries later. That other English scientific icon, Francis Bacon, believed with Plato that objects were rendered visible by emanations from the eye.

The great seventeenth-century French philosopher and mathematician René Descartes viewed light as a kind of pressure transmitted by an infinitely elastic medium that pervaded space. This was as close to a wave theory as science would come until the eighteenth century, when Christiaan Huygens would study diffraction in lenses.

CATCHING SOME RAYS

By 1672, most natural philosophers (as scientists were called in those days) believed that the speed of light was finite. Galileo even proposed a method of measuring it: two men with lanterns would stand on adjacent mountaintops. One would uncover a light and the other, as soon as he saw the first one's light, would uncover his own. The first man would then measure the elapsed time. The speed of light would then be twice the distance between the mountaintops divided by the time. Unfortunately, even if the mountaintops were 100 kilometers apart, light traveling at nearly 300,000 kilometers per second would take 0.3 millisecond to travel from one top to the other. This would make the reaction time of the second observer nearly a thousand times too slow.

In that same year of 1672, the Danish astronomer Olaus Roemer began a series of measurements of Jupiter's moon Io at an observatory near Paris. In 1610, Galileo had discovered Io and other moons of Jupiter through a new invention called the telescope. Galileo's discovery had created a sensation. It showed that one body could easily be the satellite of another, giving new credibility to the Copernican hypothesis. Some fifty years later, Newton formulated his new laws of motion. The new field of celestial mechanics was off and running.

To plot the orbit of Jupiter's moon Io, Roemer timed occultations of Io as it passed behind its Jovian companion. On each such occasion, he carefully recorded the time, extending his observations as far as the year 1675. By that time he had enough timed occultations to determine orbital periods. But when he worked out the time between successive occultations of the same moon, he discovered to his dismay that the times gradually increased over half of each year and decreased over the next half, increasing again after that.

Could Newton's new theory, the very foundation of celestial mechanics, be wrong? Was there another, unknown influence acting on Io? Or was it light itself? Roemer was well aware that in the vast theater of the solar system, all planets moved at different speeds in their orbits, the outer planets, such as Jupiter, taking much longer to go around the Sun than the inner ones, such as the Earth. In other words, it was quite possible for the Earth, as it sped around the Sun, to complete several orbits as Jupiter completed only a portion of its own orbit. Consequently, one could pretty well treat Jupiter as fixed. As it happened, the discrepancies in the orbital data followed a nearly annual cycle—a major clue.

The illustration below shows how, over a six-month period, light from Jupiter (or Io) would take longer and longer to reach Earth.

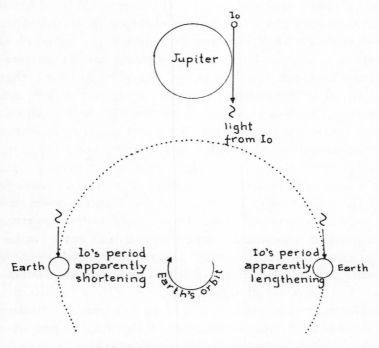

Light from Jupiter's moon crosses Earth's orbit

As Earth moved away from Jupiter, the occultations of Io would "arrive" at successively later times, while during the rest of the year, the opposite happened. The greatest difference between the two sets of data would occur between the time when Earth was farthest from Jupiter and when it was closest, a total difference of about 17 minutes. Roemer must have seen immediately that this time was precisely how long it took light from Jupiter to cross Earth's orbit.

In the 1670s, the diameter of Earth's orbit was not known precisely, the best estimates placing it at about 200,000,000 kilometers. By this reckoning, the speed of light would be

200,000,000/(17 × 60) km/sec

or roughly 196,000 kilometers per second. This value is off by approximately 33 percent.

Subsequent recalculations, based on better estimates of the diameter of Earth's orbit, improved the accuracy of this estimate slightly, but there was no real further advance until 1849, when the French physicist Armand Fizeau found the speed of light to within 4 percent of the correct value. He employed a toothed wheel and a light path that passed first through the teeth, then to a reflecting mirror some distance away, then back to the toothed wheel. Fizeau found that by driving the wheel fast enough, the return beam would strike a tooth in the wheel instead of reentering the gap it had first left. The speed of the wheel, its diameter, and the distance between teeth, as well as the length of the light path would, among them, make it possible to calculate the speed of light.

Not to be outdone by a countryman, the French physicist Jean Foucault used a rotating mirror and the same basic setup a year later. The mirror had greater sensitivity than the toothed wheel for this experiment. Foucault's value came within 1 percent of the correct speed. The subsequent history of the determination of the speed of light takes us beyond the starting point (historically speaking) of the next section. But in the years from 1924 to 1926, the American physicist Albert Michelson conducted a series of measurements between the tops of Mounts Wilson and San Antonio in California in a modern version of Galileo's original suggestion. But instead of two men with lanterns, Michelson used a variant of Foucault's rotating mirror, an eight-sided affair that, coupled with the very long light path, yielded the most accurate result then available, 299,729 kilometers per second. Michelson's measurement came within 0.02 percent of the correct value, being some 70 kilometers an hour too slow, owing to the passage of light through air. The modern value quoted above was obtained entirely in the laboratory using a relatively short light path but extremely fast electronics.

The speed of light is today considered one of the fundamental constants of nature, but not because it has been measured so accurately. It amounts to a fundamental limit on all motion, even communication, in our universe. The remarkable insights that gave birth to this fundamental knowledge had their own birth in a curly-haired boy of twelve living in Zurich.

THE ETHER

Albert Einstein was born in Ulm, Austria, in 1879, the son of a middle-class Jewish family who were then struggling to make ends meet. In his early schooling he was slow and awkward, leading at least one teacher to exclaim, "He won't amount to much." But by the time he was twelve, young Einstein had developed a strong liking for physics and mathematics. Gifted with a powerful imagination and an almost obsessive curiosity about nature and "how God works," Einstein was as much given to speculative reasoning of a specific type as he was fond of tinkering with machines that demonstrated physical principles. The "specific type" of reasoning was the famous gedanken (or thought) experiment.

At sixteen, Einstein had tried to imagine what it would be like to travel beside a ray of light, moving at nearly 300,000 kilometers per second. What would he see? His own answer (as recalled much later in life) was "a spatially oscillatory electromagnetic field at rest." This was an obvious contradiction that not even Maxwell's famous equations governing all electromagnetic phenomena (including light) could elucidate. Already Einstein could see that the only way out was to presume that somehow the laws of electromagnetism would be different for an observer at rest and one traveling at the speed of light.

Since Maxwell's equations were unlikely to be wrong, the only remaining possibility would be that nothing could travel that fast; so far had the young Einstein reasoned in his youth. To know the kind of question one might be pursuing at the end of formal schooling gave an inestimable advantage. On the other hand, Einstein's education was anything but smooth. His life in the Luitpold Gymnasium in Munich was miserable, owing to the strict Prussian atmosphere that surrounded much of the German educational system in the late nineteenth century. He hated most of his teachers and frequently displayed a rebellious attitude. In fact, he failed to graduate, being expelled. "Your presence in the class is disruptive and affects the other students," said one of his teachers.

Luckily, Einstein had been reading mathematics and physics books through his teens, steadily building competence and confidence. His father decided that Albert was better off continuing his education in Zurich. The Swiss Federal Institute of Technology required no secondary school diploma, deciding instead that admission would depend on an entrance examination. Einstein failed the exam. Although his mathematics was equal to the challenge, he had acquired none of the other subjects, such as biology. A kindly master at the institute directed the young Einstein to a cantonal school near Zurich, where he could spend the year preparing for the entrance examination. Even when he passed the exam a year later, he was still one year below the normal admission age of eighteen. He was admitted anyway.

Knowing (more or less) where he was going, Einstein studied mathematics and theoretical physics at the institute, blossoming as a student (as much as he ever would) and falling in love with Switzerland.

At the turn of the twentieth century, physics was bubbling. Dalton's atoms had proved to contain electrical charges; radioactive decay had just been observed; and Maxwell's equations described waves of all kinds, including light. The appearance of Henri Poincaré at the First International Congress of Mathematicians in Zurich at the end of Einstein's first year at the institute may have been a major influence on the young physics student. Said to be the last mathematician who would ever know "everything" there was to know about mathematics, Poincaré gave a stirring lecture, which included the following prophecy: "Absolute space, absolute time, even Euclidean geometry, are not conditions to be imposed on mechanics; one can express the facts connecting them in terms of non-Euclidean space."

Another hint of things to come was in a book by Ernst Mach called *The Science of Mechanics*. Mach was German, a physicist turned philosopher. His book stirred many students at the Zurich institute, including the young Einstein. Perhaps Mach's criticisms of Newton caught his eye. Mach railed at the notions of "absolute space" and "absolute time," urging that "relative to the fixed stars" be substituted for the former expression.

The notion of absolute space was already becoming a bit threadbare, thanks to the work of Michelson and Morley, two physicists then working at universities in Cleveland, Ohio. The wave theory of light was first proposed seriously by Christiaan Huygens in 1678, then established by elegant experiments of Fresnel and Youngs. It seemed perfectly clear that light traveled in waves, like waves on the sea. But what "sea" could be said to fill all of space so that waves of light could ripple across it at such fantastic speed? Most physicists of the late nineteenth century accepted some version or other of Descartes' suggestion that an "ether" was the basic substrate through which light traveled. Maxwell, whose equations seemed to close the book on light and other electromagnetic phenomena, also favored the idea of an ether, the "luminiferous [light-carrying] ether," as he called it.

The mysterious ether had some strange properties. First, it had to be extraordinarily rigid to transmit light at such a high speed. For example, the speed of sound in steel is seventeen times faster than the speed of sound in air because the atoms comprising the steel are in constant and intimate contact and therefore vibrate much sooner, relatively speaking, when their neighbors vibrate. Another problem with the ether was that it had to permeate material objects such as windows (since light can travel through them) and, as experiments on the speed of light in moving water clearly demonstrated, the ether must, to some extent, be dragged along by the water. On the other hand, the apparent directions of stars observed from the Earth were not affected by motion of the ether outside the Earth. Consequently, the ether outside material objects must be stationary, a fixed frame of reference that would have been heartily welcomed by the likes of Newton.

The inconsistency arose from the assumption that the ether—whatever it was—actually existed. The speed of light would obviously be constant in the ether. An observer moving through the ether would be able to measure a slowing down or speeding up of light, depending on whether the observer was moving with the ether or against it. In other words, if light was represented by a moving train and the ether by tracks, it would indeed be possible to catch up with a beam of light along a parallel track. It would be possible, as young Einstein had imagined, to watch this speeding train as if it were (relatively) motionless.

In 1882, Albert Michelson, the physicist whose measurement of the speed of light I described earlier, joined the physics department at the Case Institute of Applied Science in Cleveland. There he teamed up with a chemist, Edward Morley, from the adjoining campus of Western Reserve University. Michelson had already developed his famous interferometer for detecting the luminiferous ether while visiting Germany a few years earlier, and now, with Morley to assist, he was ready to conduct a very precise experiment.

The idea was simple. A beam of light was focused on a beam splitter, a semitransparent surface that transmits half the beam while reflecting the other half. In this manner, the emergent beam was split into two components at right angles to each other. Each beam was reflected from a distant mirror and returned nearly to the beam splitter, arriving at a screen where they could interfere with one another.

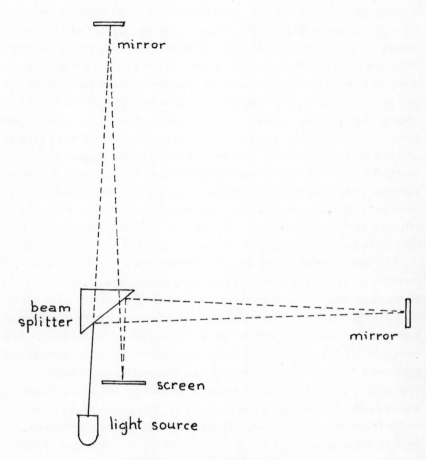

The Michelson-Morley interferometer

Since the two paths did not necessarily have precisely the same length, the two beams of light, when they rejoined, would produce a set of fringes that represented the phase difference between the two beams. If the crest of a wave in one beam met the trough of a wave in the other, the two would cancel out, producing a dark band. If two crests met, however, the waves would reinforce each other, and the result would be a white band.

The fringes or interference patterns would thus indicate, to within a wavelength of light, the exact positions of the two beams relative to one another. If one of the light paths were to lengthen by a slight amount (or if either beam were to change its relative velocity), the fringe pattern would shift by a measurable amount on the screen of the interferometer.

With the apparatus aligned so that one beam traveled parallel to the Earth's supposed motion through the ether while the other traveled at right angles to that direction, the theory behind the experiment is not difficult to understand. The beam of light traveling in the same direction as the Earth's motion through the ether would complete the first half of its journey somewhat faster than the waves traveling at right angles to the Earth's motion. On the return journey, the waves would travel somewhat slower. Both halves of the path taken by the right-angled beam would be traversed at the same speed. If the ether were real, the beam traveling parallel to the Earth's motion through the ether would take less time to travel the "downstream" leg of its journey than it would to return. The beam traveling at right angles to the Earth's motion through the ether, however, would take the same amount of time on both paths.

The average speed of the beam traveling parallel to the ether would not exactly equal the speed of the other beam, however. The whole experiment hinged on the solution of an old elementary school problem.

I will follow the time-honored tradition of invoking two swimmers in a river, calling them "A" and "B." Both A and B dive into the middle of a river from a bridge at the same time and swim at exactly the same speed relative to the water. Swimmer A swims 100 meters downstream, then 100 meters back to his starting point on a bridge. Swimmer B, meanwhile, swims 100 meters across the river at right angles to the direction of flow, then back again. Assuming that the rate of flow is everywhere the same in the river, which swimmer arrives back at the starting point first?

Suppose the current travels at 6 meters a minute and that both swimmers travel at the rate of 30 meters a minute. Swimmer A travels the downstream leg of his journey at the speed of 36 meters a minute relative to the land, taking just $100/36 = 2.78$ minutes. On the return leg, he travels at 24 meters per minute and therefore requires $100/24 = 4.17$ minutes, for a total travel time of 6.95 minutes. Swimmer B, on the other hand, completes both legs of his journey in the same time of $100/30 = 3.33$ minutes, for a total travel time of 6.66 minutes. Clearly, swimmer A takes longer.

Michelson and Morley set up the interferometer in what they thought might be the direction of the Earth's travel through the ether, but found that the interference fringe did not shift in their instrument. Well, perhaps, they hadn't guessed right, so they tried another angle. That produced no effect, either. They tried every angle they could think of, even tilting the interferometer toward the ceiling. Still no effect. They concluded that there was no ether, at least not one with the properties attributed to it. In short, light was not propagated through any mysterious medium. Moreover, the speed seemed to be the same in all directions!

They published their results, and the effect on physics was profound, creating a first-class crisis from which no one could see a way out. Maxwell's equation required an ether (or some such medium) that could propagate light waves through it. Yet the ether had not the slightest effect on the speed of light waves. It was this second observation that caught young Einstein's eye. Could the speed of light be constant and independent of the speed of the observer? How could that be?

Einstein completed his degree at the Zurich institute in 1900 but, owing to uneasy relations with one of his professors, did not get the equivalent of a postgraduate position. Thus he was thrown into a patchwork academic life of short-term teaching posts while he worked on his doctoral thesis on the kinetic theory of gases. By 1902 he had finished the thesis and had found a job at the Swiss patent office, in which he held the post of technical expert, charged with examining patents for inventions that depended on tricky physical effects. He would stay with the patent office for seven years. It was not until 1905 that he defended his thesis at the Zurich institute, an examination that was "touch and go," according to one biographer. In that same year Einstein published three papers in the journal *Annalen der Physik*. One paper linked Brownian motion with the existence of molecules. The second, more remarkable paper, explained the photoelectric effect; and the third, most remarkable paper of all, outlined the principles of special relativity.

Physicists had noticed that when a beam of light, even a weak beam, was directed onto a metal surface, it would trigger a flow of electrons from the plate, a phenomenon known as the photoelectric effect. The energy of the electrons thus emitted, moreover, did not depend on the intensity of light, only on its color, or wavelength. This particular phenomenon could not be predicted by Maxwell's equations. In his paper Einstein derived equations of great generality that enabled him not only to declare that the radiation emitted by certain heated bodies must consist of individual particles of light, but also to explain the photoelectric effect using the new theory. What the German physicist Max Planck had hypothesized only five years earlier—that under some circumstances, light was best considered as occurring in tiny packets called quanta—Einstein demonstrated to be real. The photon was born, or, should we say, discovered. Remarkably, physics had traveled full circle, from the Pythagorean view of light as a stream of particles (reinforced by Newton) to the discovery of the wave nature of light by Huygens and Fresnel, neatly packaged in the wave equations of Maxwell, and back to the "new" view of Planck and Einstein that light consisted of particles after all.

Einstein was thus among the founders of a new branch of physics to be known as quantum mechanics. He could not foresee the distress it would eventually bring to his intellectual life, causing the now famous complaint "God does not play dice with the universe."

THE THIRD PAPER

From age sixteen, Einstein had been pondering the nature of physical reality, always returning to light. His strange and contradictory dream of the stationary light wave never left him. It blossomed into a problem he would view from every conceivable angle during his undergraduate and graduate days. Even as he labored on molecular dynamics for his dissertation, he would interrupt himself occasionally as a new thought struck him.

He was alert to clues, wherever they might come, not only to the remarks of Poincaré about non-Euclidean geometry or Mach's attempt to abolish absolute space and time, but also to some strange new developments in Ireland, Holland, and France.

George Fitzgerald was a professor of "natural and experimental philosophy" at Trinity College in Dublin. Fitzgerald had been troubled by the very same problem as the young Einstein, but from a different angle, so to speak. According to the outcome of the Michelson-Morley experiment, light always traveled at the same speed, regardless of the state of motion of its source. The light from two stars, one speeding away from the Earth, the other speeding toward it, would arrive at precisely the same speed, nearly 300,000 kilometers per second. He did not believe that the velocity of light could be unaffected by the motion of its source. The only escape from this logical cul-de-sac, as far as Fitzgerald was concerned, was to suppose that any object in a state of motion was subject to a contraction in the direction of its motion. In other words, a single wave from the approaching star would appear to shorten in the direction of its travel so that its wavelength would be shortened (toward the blue end of the spectrum) when measured. Similarly, light from a receding source would be shifted toward the red and in this way, the motion of the source would directly influence the physical properties of the arriving waves.

The premise was questioned by other physicists, as the hypothesis was untestable: a meter stick set up to measure the effect of high velocity on a moving object would have to travel with it. Aligned in the direction of the object's motion, the meter stick would contract by the same amount and give precisely the same reading as when the object was stationary. Perhaps the following little limerick, recited in a course on relativity theory I once attended, dates back to the days of derision:

> There once was a swordsman named Fisk
> Whose thrust was exceedingly brisk.
> So quick was his action, the Fitzgerald contraction
> Shortened his sword to a disk!

The actual transformation in length developed by Fitzgerald had a relatively simple algebraic form. If an object traveled at velocity v in a given direction, its length would shorten by the ratio r:

$$r = \sqrt{1 - (v^2/c^2)}.$$

The shortening is practically unnoticeable at the velocities we encounter in ordinary life. For example, suppose you're driving a car at 100 km/hr (about 60 mph); the actual amount by which the car—and you—shorten from the point of view of a stationary observer can be readily calculated from the formula. Given that 100 km/hr is approximately 0.028 km/sec, we can form the ratio

$$v/c = 0.028/299{,}792$$
$$= 0.093 \times 10^{-6}.$$

If we square this number, we get

$v^2/c^2 = 0.859 \times 10^{-8}$.

For the next step in the calculation, we must subtract this amount from 1 and then take the square root of the result:

$1 - 0.859 \times 10^{-8} = 0.999999991$.

The square root of this quantity is 0.999999995, and the length of your car (say, exactly 4 meters) has shortened to

4×0.999999991 meters.

At 100 km/hr, your car is going nowhere near the speed of light, so it is not surprising that the actual shrinkage is likely to be small. In this case, it becomes

$4.0 - 3.999999964 = 0.000000036$

or 0.036 micron, 1 micron being the diameter of a smallish bacterium.

As for the swordsman Fisk, if he should manage to thrust at just half the speed of light, his sword would still be somewhat more effective than a disk; it would shorten by a mere 13.4 percent.

To see how Fitzgerald derived his contraction ratio, we can almost follow his mental footsteps. He would, of course, have been thinking about the famous Michelson-Morley experiment, since that was the source of all the trouble. As you may recall, the interferometer had two arms with a mirror at the end of each arm. One arm (the "ether arm") would supposedly be parallel to the flow of the ether as the Earth moved through it. The other arm would be at right angles to the flow. The contraction that Fitzgerald hypothesized would take place in the ether arm, exactly enough to shorten it to the right degree. Light, which had to fight its way upstream against the ether, would, of course, be traveling more slowly, but the ether arm would contract by exactly the right amount to make the time taken by the light beam along that arm equal to the time taken by the other beam. The following figure shows the apparatus once again, but this time with an outside observer.

Fitzgerald examined everything from the point of view of an observer who was stationary with respect to the ether. To such an observer, Michelson, Morley, and their lab might go speeding past his or her position, but as they sped by, this privileged observer would be able to deter-

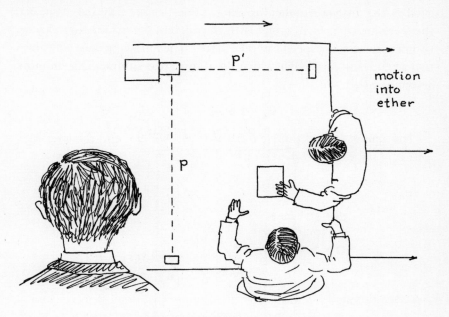

Michelson and Morley being watched

mine the exact length of the path along the ether arm. In the diagram I have labeled it as P', whereas the perpendicular path has length P. To the observer, P' would appear shorter than P.

Suppose that the ether arm of the interferometer was pointing "upstream" into the ether. Then the time taken by light to travel the distance P' to the distant mirror would be

$$P'/(c-v),$$

since light (c) was slowed by the passing ether (v).

After being reflected in the mirror at the end of the ether arm, the light would now travel the downstream leg of its path, traveling a little faster than before and therefore taking somewhat less time:

$$P'/(c+v).$$

To determine the time taken by light to travel the path of length P, Fitzgerald may have reasoned by analogy; let's return to the swimmer. To travel across the river at right angles to the direction of flow, the swimmer must angle somewhat upstream so he does not drift steadily downstream, even as he crosses the river. Traveling at the speed of light (so to speak), the swimmer makes an angle with the current (or ether) that exactly compensates for the speed v of the current. The following diagram shows that a right-angled triangle is just the thing to determine the swimmer's resultant speed.

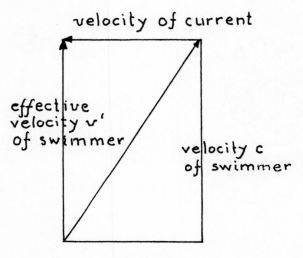

Diagram of velocities

Pythagoras' theorem has to be the most useful result in all of mathematics. Here it tells us that the square of c must equal the square of v plus the square of the unknown (cross-stream) velocity v'. The latter velocity must therefore be

$$v'^2 = c^2 - v^2.$$

The time it takes light to traverse the path of length P must then be:

$$\frac{2P}{v'} \quad \text{or} \quad \frac{2P}{\sqrt{c^2 - v^2}}.$$

Since the sum of the two traversal times taken by light along the ether arm of the apparatus must equal the time just derived, we have, finally,

$$\frac{P'}{(c-v)} + \frac{P'}{(c+v)} = \frac{2P}{c^2 - v^2}.$$

A little algebra comes in handy at this point. To simplify the expression on the left-hand side of the equation, we may first take out the common factor P' and concentrate on the expression

$$1/(c - v) + 1/(c + v).$$

To bring both fractions to a common denominator, we simply use the product of the two denominators and write

$$\frac{c + v}{c^2 - v^2} + \frac{c - v}{c^2 - v^2}.$$

The terms with v in them cancel out and we are left with

$$\frac{2c}{c^2 - v^2}.$$

Finally, we multiply by the P', which we had left out of the expression for the time being:

$$\frac{2cP'}{c^2 - v^2}.$$

We are, as some mathematicians say, almost home. The equation that we started with can now be written

$$\frac{2cP'}{c^2 - v^2} = \frac{2P}{c^2 - v^2}.$$

In the following steps I form the ratio P'/P, which is identical to r. But first I multiply both sides by the square root of the expression $c^2 - v^2$, which leaves us with

$$\frac{2cP'}{\sqrt{c^2 - v^2}} = 2P.$$

The ratio

$$\frac{P'}{P} = (\sqrt{c^2 - v^2})/c$$

can be further simplified by taking the factor $1/c$ inside the square root, where it becomes $1/c^2$:

$$\frac{P'}{P} = \sqrt{1 - v^2/c^2}.$$

There is a strange irony in this formula. In what would turn out to be a vain attempt to save the ether, Fitzgerald had invoked a mysterious contraction of the ether when, in reality, it was everything *else* that was contracting. And yet it was exactly the right expression for a profound overhaul of our ideas about space, time, matter, and energy.

In the meantime, physicists regarded Fitzgerald's notion of a contracting ether as far-fetched, to say the least. In 1894 Fitzgerald wrote his friend Hendrik Lorentz, a well-known Dutch physicist, "I have been rather laughed at for my view over here."

In spite of the barely plausible nature of Fitzgerald's contraction, Lorentz saw something more in it. Such a contraction might result when a body moved through the ether. The motion might disturb the equilibrium of electrical charges composing the body, and its particles would change their relative positions to assume a new shape. Lorentz, one of the discoverers of the electron, gave the hypothesis a new respectability as the nineteenth century drew to a close. At the same time, he realized that it would play havoc with ordinary notions of time and space, at least in the sense that ordinary addition or velocities, even calculations of position, would all have to be recast to allow for what were then coming to be known as the Fitzgerald-Lorentz transformations. His worries foreshadowed the world of relativity, which, at that time, was hardly more than a cloud of questions and notions in the brain of young Einstein, still a student in Zurich.

The great Henri Poincaré, meanwhile, felt that deeper issues were afoot. He addressed a symposium on the achievements of science in the century just past in St. Louis, Missouri, in 1904, "Perhaps we should construct a whole new mechanics, of which we only succeed in catching a glimpse, where, inertia increasing with the velocity, the velocity of light would become an impassable limit."

THE GREAT INSIGHT

For Lorentz and Fitzgerald, the transformation equations amounted to a property of bodies, specifically bodies that contained electrons (just about all forms of matter). Lorentz had added a further contraction, one in time itself, so they would fit with Maxwell's equations, which described electromagnetic waves in both space and time. The equations,

which related two frames of reference—one "fixed," the other moving—
were simple enough to be understood by a high school student:

$$x' = \frac{x - vt}{\sqrt{1 - v^2/c^2}}$$

$$y' = y$$

$$z' = z$$

$$t' = \frac{t - \frac{v}{c^2}x}{\sqrt{1 - v^2/c^2}}.$$

This version of the Lorentz-Fitzgerald transformations is not the
most general one. It assumes that one frame, with coordinates x, y, and z,
is stationary, while the second frame, with coordinates x', y', and z', is
moving at a velocity v relative to the first frame in the x direction alone.
In the most general version, the equations for y' and z' would look similar to those for x'.

Although Einstein used these very transformations in his groundbreaking 1905 paper introducing the world to the special theory of relativity, he interpreted them somewhat differently. For Einstein, the
transformations were a property of space (rather, space-time) itself and
in no way depended on the properties of electrons. He had already seen
that many physical measurements produced the same result, no matter
what state of uniform motion the laboratory happened to be in. For
example, the Michelson-Morley experiment produced a null result in
October as well as in April, six months later, when the Earth was moving
in the opposite direction in its annual course about the Sun. There was
simply no experiment a person could perform that could tell what "ultimate" state of motion one was in.

The transformations applied, moreover, to all frames of reference, no
matter what their state of motion, as long as velocity was a constant.
Such frames of reference are called "Galilean frames" in honor of the
Italian physicist Galileo Galilei. To qualify as Galilean, they could be neither rotating nor accelerating. Thus it was easy to determine whether
one's frame of reference happened to be Galilean or not. Neither form of
motion is relative in the sense that if your laboratory were spinning,
objects would tend to slide off tables. If accelerating, you would feel a
force pulling you toward one wall.

The following figure shows two Galilean frames, one with its axes

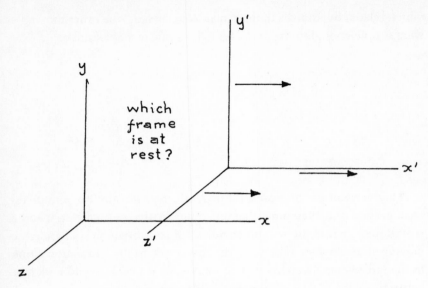

Galilean frames of reference

labeled x, y, and z, the other with axes labeled x', y', and z'. Let us suppose that you, the reader, happen to be moving with the first frame (x, y, z), while I, the author, happen to be moving with the second one (x', y', z').

As a courtesy, I have left your Galilean frame apparently at rest, while mine appears to be moving to the right at velocity v. Of course, to me your frame appears to be moving in the opposite direction at the same speed. As far as Einstein was concerned, that would be about all we could say concerning our motions. The transformation equations enable me to determine my coordinates in terms of yours. If I want to know the x' coordinate of a point within my own frame, for example, I take the corresponding point in your frame and subject it to the appropriate formula, the one involving x, x', and v. Thanks to the transformation equations, everything going on in your frame of reference can be expressed in my own coordinates. I can measure distances in your frame of reference, for example.

As a purely practical matter, however, I will have to equip each of us with a meter stick and a clock within our respective Galilean frames. To overcome certain difficulties pertaining to our high speed relative to one another, and our consequent separation, we also will have to use telescopes with which to view each other's instruments. To me your clock will appear to be running slow, and your ruler also would appear short. My clock also would appear equally slow to you and my ruler equally short. For example, if $v = 59{,}958{,}492$ (one-fifth the speed of light in meters per second), our respective clocks would each appear to the other to run at

$$\sqrt{1 - 0.04} = 0.98$$

seconds per second, so to speak. In other words, your clock would appear to lose 2 out of every 100 seconds when compared to mine, while my clock would appear equally faulty to you. Our meter sticks also would appear shorter to each other, by about 2 centimeters in both cases. Admittedly, such discrepancies are not very dramatic. But at four-fifths the speed of light, the Lorentz-Fitzgerald transformations produce more impressive differences. In this case the transformations yield a much smaller ratio:

$$\sqrt{1 - 0.64} = 0.60.$$

At this relative velocity, each of us will observe the other's clock to lose 40 seconds out of every 100. Also, each of us will see the other's meter stick shorten to 60 centimeters—although none of the marks on these rulers will disappear. They will just look closer together to the other observer.

Besides reinterpreting the Lorentz-Fitzgerald transformations anew, Einstein saw important consequences for the observed behavior of mass. For example, prior to Einstein, physicists had always determined the kinetic energy of a moving object of mass m by using the formula

$$mv^2/2.$$

Einstein, applying the Lorentz-Fitzgerald transformations, obtained a new formula, with some startling implications:

$$\frac{mc^2}{\sqrt{1 - v^2/c^2}}.$$

One of these implications concerned the behavior of an object traveling ever closer to the speed of light. As v approaches c in value, the expression inside the square root approaches 0. Division by a quantity that approaches 0 produces a quantity that approaches infinity. In other words, the object's kinetic energy, as observed by us, would simply become greater and greater without limit. At $v = c$ the energy would appear infinite—an utter impossibility, according to Einstein. Another important implication concerned the value of the formula when the object is standing still relative to the observer's frame of reference. In this case, the value inside the square root is 0 and the energy of the object, called its "rest energy," is mc^2. But that is another story.

IS THERE A WAY AROUND IT?

As everyone knows, a popular way of measuring distance in deep space uses a unit called the light-year. The nearest star to our solar system, Proxima Centauri, is about 4.3 light-years away, but the vast majority of stars in our galaxy are thousands of light-years away. Traveling at the speed of light, it would take you as many years to reach these stars. Science fiction writers, impatient with this annoying limitation, have invented a hyperspace drive that circumvents the problem. The idea is to "fold" space back upon itself so the region I inhabit actually comes into contact with a region thousands of light-years away. I pilot my spaceship across the fold, and voilà! I'm there.

Such travel would not violate the speed-of-light edict, since the body thus translocated would not have been moving in the classical sense. At the same time, folding space, even if it were possible, might well involve the expenditure of prodigious amounts of energy, making it never more than a hypothetical possibility.

The speed of light not only limits the velocity of any classical (large) body but also limits the speed at which information can travel. Some have argued that quantum effects involving instantaneous changes in the state of two widely separated photons would provide the basis for a faster-than-light signaling system. Unfortunately, as we will discover in the next chapter, there is no way to manipulate the state of these photons, so even if the effect is instantaneous, no information can be transmitted faster than light.

Perhaps the speed of light is not constant, after all. Would that change its nature as a cosmic speed limit? In the early 1980s two Australian physicists, Barry Setterfield and Trevor Norman, published a claim that the speed of light was not constant. They made the case that even over the past 400 years there had been a small but statistically significant decline in the velocity of light. They theorized that the speed of light shortly after the big bang had been many orders of magnitude greater, that it had declined exponentially since then, and was now dropping by nearly imperceptible amounts every century.

The recognition that c is not constant after all has gained ground in the past decade. Does this not put Einstein out the window? In fact, neither the special nor the general theory of relativity depend on the speed of light being a fixed constant, as long as it is the same for all observers, which, apparently, it is.

Perhaps the best way to test the universal speed limit (even if the cosmic police are steadily reducing it) is to ask what would happen if one could go faster than light. According to special relativity, both space and time would shrink by the Fitzgerald contraction:

$$\frac{1}{\sqrt{1 - v^2/c^2}}.$$

At such a juncture, things would become distinctly creepy. The quantity inside the square root would become negative because the term v^2/c^2 would exceed 1. Distance and velocity would become imaginary numbers. What does that mean? I haven't a clue, and I shudder to think.

· 3 ·

The Quantum Curtain

Unknowable Particles

⟦ THE DETAILED BEHAVIOR OF ANY QUANTUM
SYSTEM, WHETHER IT CONSISTS OF ELECTRONS,
PHOTONS, OR ATOMIC PARTICLES, CANNOT BE
PREDICTED BY ANY MATHEMATICAL LAW OR
COMPUTER PROGRAM. ⟧

IF THE EARLY DECADES of the twentieth century were decisive for
modern mathematics, no less is true of modern physics. Einstein's theory
of relativity was revolutionary enough, with its strange amalgamation of
space and time and its counterintuitive ideas, such as light bending in
gravitational fields. But the rise of quantum mechanics signals an even
stranger development, with an interpretation that Einstein himself resis-
ted to the end of his days.

The theory of relativity is shelved within "classical" physics because
its equations demand a classical notion of reality. A body traveling

through space always has a definite position and velocity, and we can measure these quantities as accurately as we like, independently of each other. Even if the body is traveling at close to the speed of light and its shape changes as a result, we can predict the degree of distortion.

In quantum mechanics, all that goes out the window. For example, we can measure the position or velocity of an electron, but, as we will see shortly, we cannot do both simultaneously. Moreover, we cannot be sure that these quantities have *any* value unless and until they are measured! Yet the success of quantum mechanics, as measured by the accuracy of its predictions, is unparalleled in physics.

The uncertainty about the position, momentum, spin, and other attributes of a quantum particle may have its roots in what appears to be a random element operating at the very basis of quantum phenomena. To the classical physicist the complaint of Einstein that "God does not play at dice" expresses the ultimate dismay at the loss of causality and determinism. I do not propose to rescue Einstein from his dilemma, but he might just have found comfort in the following proposal: Whatever God is doing, He's doing it behind what I call the quantum curtain, a kind of reality veil behind which we cannot see. In other words, whether the behavior of quantum particles is random or not, we may never know the difference. We will not, in any case, ever be able to predict the detailed behavior of such systems.

The roots of quantum mechanics penetrate well back into the nineteenth century, with the discovery of discrete atomic spectra by the Swedish physicist Anders Angström and the Swiss mathematician Jacob Balmer, but it was Einstein himself who opened this Pandora's box when he published his findings on the photoelectric effect in 1905. I will return to the history for the insight it affords, but I cannot resist rushing to the central phenomenon.

THE TWO-SLIT EXPERIMENT

The laboratory is dark. Somewhere a vacuum pump wheezes. A physicist stares intently at a screen, where he sees minute flashes of light several times a second. These are individual photons striking the screen after passing through two slits in an opaque shield, as shown in the figure. The photons come from a very weak light source, well behind the screen and aimed at it. Two photodetectors tell the physicist which of the slits each photon passes through. At the moment, the detectors have been turned off.

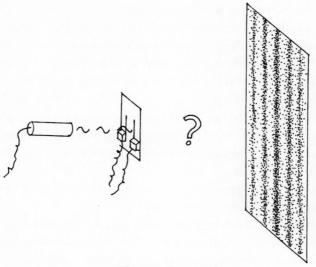

The two-slit experiment

The pattern on the screen demonstrates the well-known phenomenon of wave interference. Light passing through the slits interferes with itself to produce alternating dark and light bands. Of course, the physicist does not see the interference pattern in this case because the photons arrive at the screen too slowly to make out any kind of pattern. But if he replaces the screen by a photographic plate, any underlying pattern in the behavior of the photons will show up when the plate is developed.

If the physicist turns on the photodetectors, he can immediately tell which slit a given photon passed through. Strangely, even though he knows which slit each photon passed through, no theory will tell him in advance which slit a given photon will use. It is completely random as far as he or anyone involved in the study of quantum mechanics knows. With the light source aimed just so, he can even arrange that the probability is 50–50. Half the time a photon passes through one of the slits, half the time the other. If he tosses a coin, he'll have a pretty fair approximation to the behavior of these photons. The physicist can guess, of course, and approximately half the time he will be right. But neither he nor any other scientist can predict what an individual photon might do. Whatever process might be thought to lie behind the deployment of the photons is hidden from his view behind the quantum curtain.

Life in the quantum lab gets considerably creepier than this. If the physicist develops the plate after an experimental run when there are no detectors beside the slits, the pattern portrayed in the figure above shows up on the plate. Obviously the photons were interfering with each other.

But wait! How can the photons interfere with each other when they pass through the slits one at a time? It turns out that each photon may be considered as a wave. It interferes with itself. But to interfere with itself, it must pass through both slits. This view is confirmed by what happens when the physicist runs the experiment with the detectors turned on. This time the developed plate shows no pattern at all, just a random scattering of points on the plate.

It appears that if the physicist does not measure which slit the photon passes through, it passes through both.

We're not out of the creepy department yet. It may be that before it "chooses" which slit to pass through, the photon has no real existence. It also may be that the consciousness of the experimenter plays a crucial role in the experiment.

In what follows we will witness other experiments and wander a historical path that leads from Einstein to the Danish physicist Niels Bohr and on to Werner Heisenberg, John von Neumann, and other luminaries of the quantum universe.

ROOTS OF THE DISCOVERY

Einstein has no one to blame but himself (although the German physicist Max Planck and the French physicist Louis de Broglie must share some of the blame) for the scandalous behavior of light. It was Einstein who gave the first reasonable interpretation of an experiment performed by the German physicist Philipp von Lenard and others around the turn of the nineteenth century. Von Lenard and his colleagues had discovered that if light is shone on a metal sheet, electrons are ejected from the metal. The puzzling thing was that light, as everyone knew, was a wave and that it possessed energy. Therefore, the more energy he put into the light beam in the form of more photons, the higher the expected velocity of ejected electrons. But this is not what happened. Instead, the speed of ejected electrons was the same, but more electrons left the metal. On the other hand, if the *color* of light was changed, so was the *speed* of electrons leaving the metal. The higher the frequency of the light beam, the higher the velocity of the departing electrons.

The key to understanding these phenomena had been provided by Planck in the year 1900. Einstein inserted the key and opened the door to the quantum age. Planck had spent arduous years working on one problem: Why did the classical wave theory of Maxwell fail to solve the so-called black body problem? "Black" is the color of a conceptual cavity in an unheated body of unspecified material. As the material is heated, the walls of the cavity slowly turn from black to red. Higher temperatures produce further changes in color, to orange, then yellow, until, at a high enough temperature, the body glows white. Thus the higher the energy contained in a black body, the higher the frequency of emitted light would be. Assuming that particles composing the black body were free to have any energy, Maxwell's theory predicted that black bodies should glow bright blue at all temperatures! The reasons for this outcome were technical, and Planck thought he saw a way around the difficulty. By allowing the particles that composed the black body to emit or absorb energy of one specific frequency, the equations gave the desired result; as the body was heated to a certain temperature, the color corresponding to that temperature emerged.

Planck had solved the black body problem in a way that made his fellow physicists suspicious. Surely, to assume that energy came in small, discrete packets was something of a "kludge." Behind the questionable solution, moreover, there lurked an implicit law that required energy to be quantized.

The conclusion based on Maxwell's laws was mistaken not because Maxwell's laws were wrong but because physicists had assumed that the particles making up the black body were free to have any energy whatever. Planck's new law stated that the energy of the particles was constrained by the rule

$$e = nhf,$$

where e is the energy of a particle, h is a constant, and f is the frequency with which the particle vibrates. The symbol n stood for a positive integer, which could take any value 1, 2, 3, and so on. This formula required that the energy of a particle in the black body could only have values that were multiples of h*f*, in effect. Energy occurred in discrete packets, or quanta. There was, moreover, a specific value of h that gave the formula the power to describe the black body effect perfectly, namely

h = 4.14 × 10^{-15}.

In spite of the fact that the new theory explained the color change in a body undergoing heating, the scientific world ignored Planck's bold hypothesis until 1905. In that year, Einstein published three earth-shaking papers, including his explanation of the photoelectric effect based on Planck's idea. In the process, Einstein explained the puzzling phenomena noted by von Lenard and others by reviving the Newtonian idea of corpuscular light. If monochromatic light could be considered as a stream of particles, each with the same frequency, then each particle of light would eject one electron from the metal's surface. This explained why more intense light only resulted in more electrons being ejected from the metal. According to Planck, the energy of such a particle was proportional to its frequency. Thus only electrons kicked out of the metal by light particles of higher frequency would have higher energies.

The "particles of light" would soon become known as "photons," as if naming the corpuscle would tame the puzzling dual existence it had introduced to physics. Photons, like all other fundamental particles, have both wave and particle qualities, depending on how you look at them, so to speak. Their double nature is referred to as wave/particle duality.

How could light consist of both particles and waves at the same time? To this day, physics has no good answer to this question. But soon, thanks to a Danish physicist, the question would become meaningless.

NIELS BOHR AND THE COPENHAGEN SCHOOL

From 1905 to 1925, physics was in ferment, and not just over the new wave-particle dichotomy introduced by Einstein and others. The French physicist Louis de Broglie had declared that not only did waves have particle qualities but also all particles would be found to have wave properties, such as frequency. Very soon, the wavelength of the electron was measured by the American physicists Clinton Davisson and Lester Germer.

In the preceding "classical" era, physics had been dominated by the concepts of matter and field, complementary but very distinct ideas. The emerging physical reality blurred the distinction. If matter, consisting of particles and fields, were the stuff of waves, then matter and field could no longer be considered distinct and separable entities. The stuff of which the universe was made could be both simultaneously.

But what was that stuff? Quantumstuff?

Niels Bohr was born in 1885, the very year that a humble teacher of mathematics in a Swiss girls' school discovered the mathematical law behind the hydrogen spectrum. Jacob Balmer had inferred, purely on the basis of spectral data published by the Swedish physicist Anders Angström, the magic formula that described the spectrum of the hydrogen atom.

It was perhaps the discrete lines of the hydrogen spectrum that first hinted at the discrete nature of atomic reality that scientists, including Bohr, were destined to wrestle with. It was Bohr who would eventually explain these lines as due to specific, quantized energy levels in the hydrogen atom. When a hydrogen atom received energy, an electron would be bumped up to a higher energy level, reradiating the energy as a photon if it dropped back to its former state. The frequency of the emitted photon depended on the energy level occupied by the electron that emitted it, higher frequencies implying higher energies. Each line of the hydrogen spectrum therefore reflected a different energy level that an electron might occupy within the hydrogen atom.

Bohr's best work began when he was relatively young and working in the laboratory of Ernest Rutherford in Manchester. The problem of the day had been stumbled upon by Rutherford himself, who had recently discovered that atoms were mostly empty space, with a tiny concentration of positive charge at its center and one or more electrons forming the "bulk" of the atom's physical dimensions. Rutherford wondered whether the electrons might orbit the nucleus like miniature planets. The problem with this view was that in circling the atom, an electron that was subject only to the classical laws of electromagnetism ought to be radiating energy as it circled the atom. But if the electron lost energy continuously, it ought to spiral into the nucleus, never to be heard from again.

Bohr solved the problem by turning to Planck's quantum in what would be only the third major application of Planck's ideas. By postulating that electrons could only have certain, discrete energies, a continuous loss of energy was prohibited, and orbits were automatically stable. The very word "orbit," however, reflects this early and tentative planetary model of the atom. Later, when it was realized that electrons did not necessarily enjoy a particlelike existence within atoms, physicists replaced the unfortunate word by "orbital" (good) and "energy level" (better).

Furthermore, when Bohr used Planck's constant to derive the diameter of the hydrogen atom—bingo! The numbers as derived and the numbers as measured matched closely. Bohr's theory also correctly predicted the energy levels of the hydrogen atom and, consequently, the positions of the lines in the hydrogen spectrum. A puzzle exactly as old as Bohr himself was thus solved in 1913, the year his paper "On the Constitution of Atoms and Molecules" appeared.

MATHEMATICS OF THE QUANTUM

An electron, photon, or other fundamental constituent of our world has a fundamentally mysterious existence. Rather than try to plumb the "reality" of these entities, quantum mechanics consists merely of rules for manipulating their attributes. What are those attributes? Briefly, we can divide them into static and dynamic. Static attributes such as charge and mass are always the same for a given type of particle. Dynamic attributes such as position, velocity, and spin may be different each time they are measured. Quantum mechanics gives rules for predicting the latter values in a statistical sense. It might predict that half the photons in the two-slit experiment will pass through slit A without being able to say which photons, in particular, would do so.

An early and very important development of quantum mechanics helped to explain the anomalous behavior of photons and other fundamental particles. In 1925 the Austrian physicist Werner Heisenberg published his famous uncertainty principle, which says, in effect, that one can measure the position of a fundamental particle, or its momentum, as accurately as one likes, but one can never measure both.

We can illustrate the physical reality of Heisenberg's principle by conducting a one-slit experiment: let a stream of photons pass through a narrow slit and you will see a diffraction pattern, with individual photons materializing as points of light that collectively build up the pattern shown in the figure.

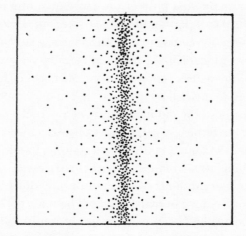

Diffraction pattern for photons passing through a single slit

If we now narrow the slit, making the photon's lateral position more certain, the pattern widens. Evidently the lateral momentum of these photons has undergone some kind of change. Some of the photons passing through the slit now diverge horizontally to a much greater degree, even as others pursue something much closer to their trajectories in the earlier experiment. Statistically speaking, even if we are more certain about the horizontal position of each photon as it passes through the slit, we are less certain about its lateral momentum. The photons that appear much further from the center of the pattern in the second experiment have greater lateral momenta than those in the first experiment. In general, we have paid a price for our improved certainty about the horizontal position of each photon in being less certain of its momentum. This relationship between momentum and position for photons has been found to hold for all quantum entities.

Heisenberg's uncertainty principle is expressed by a formula that relates our ability to measure a particle's momentum to our ability to measure its position. If we denote our uncertainty about position by Δx

and our uncertainty about momentum by Δp, the principle is readily formulated:

$$\Delta x \cdot \Delta p \geq h,$$

where h is Planck's constant.

Heisenberg was the first physicist to place quantum mechanics on a solid mathematical foundation. In 1925 Heisenberg, along with German physicist Max Born, developed matrix mechanics, a system for representing attributes of a fundamental particle or atom by a matrix of numbers. Each number of the matrix represented the quantum difference between two energy levels, and Heisenberg discovered that to multiply two quantities, such as the position and momentum of a particle, he had to multiply the matrices together. In 1925 most physicists were competent in school mathematics but frequently lacked training in more sophisticated forms. Heisenberg was unfamiliar with matrices. It was Born who pointed out that the products of these multiplications of Heisenberg's reminded him of a course in matrix algebra he once took. It turned out that mathematicians had been using matrices, multiplying and adding them for approximately 100 years. In particular the two remarked on the fact that in general the product of two such matrices was not commutative. In other words, while we may form the product of two ordinary numbers such as 3 and 7, we may multiply them in any order we like:

$$3 \times 7 = 7 \times 3.$$

Not so with matrices. Here, for example, are two matrices:

$$\begin{pmatrix} 2 & 6 \\ 4 & 2 \end{pmatrix} \cdot \begin{pmatrix} 1 & 4 \\ 7 & 3 \end{pmatrix} = \begin{pmatrix} 44 & 26 \\ 18 & 22 \end{pmatrix}.$$

To form a product like the one shown, one multiplies rows of the first matrix by columns of the second matrix. The entry in the ith row and jth column of the product is thus formed by multiplying the ith row of the first matrix by the jth column of the second one. The entry in the first row and second column of the product above is 26, obtained by multiplying the first row of the first matrix, namely (2 6), by the second column of the second matrix, namely (4 3). The product is taken pairwise: $2 \times 4 + 6 \times 3 = 8 + 18 = 26$.

If we now reverse the order of these matrices, we get something a little different on the right side:

$$\begin{pmatrix} 1 & 4 \\ 7 & 3 \end{pmatrix} \cdot \begin{pmatrix} 2 & 6 \\ 4 & 2 \end{pmatrix} = \begin{pmatrix} 18 & 14 \\ 26 & 48 \end{pmatrix}.$$

Mathematical operations in which the order of multiplication makes a difference are called noncommutative. Matrix multiplication is a noncommutative operation. Strangely enough, it was the failure of his matrices to commute that set Heisenberg on the path to discovering his famous uncertainty relation. The product of two matrices for the position and momentum of a photon, for example, is the result of measuring these attributes in that order. Measure them in the opposite order and you get a different result. To Heisenberg, this meant that measuring the position of a photon meant destroying our knowledge of its momentum—and conversely.

The year 1925 was something of an *annus mirabilis* in physics. For in the year of Heisenberg's discovery, two other physicists formulated their own versions of quantum theory. The German physicist Erwin Schrödinger developed an all-wave formulation of quantum mechanics with a single equation, today called the Schrödinger wave equation. Meanwhile, in England, Paul Dirac formulated his "transformation theory," a representation of quantum particles as vectors in a high-dimensional space. Dirac developed a technique for transforming from one coordinate system to another; hence the name of his theory.

By the end of 1925, physics had three completely different-looking theories of the quantum. It was Dirac who, with brilliant insight, discovered that the other two theories could be formulated as special cases of his own, depending on what sort of coordinate system he chose for his system. This surprising development echoes the equivalence among the three separate formulations of computation explained in chapter 7, namely the theories of Church, Turing, and Kleene. In this case, however, the explanation lies closer at hand: each was trying to express the same quantum facts with his theory—reading from the same script, in effect. In formulating what it means to compute, however, each of the latter researchers probably had quite different ways of imagining a computation. For one thing, in those days there were no computers.

Today, quantum mechanics is an essentially mathematical subject that can be described in a nutshell only for those who know a little about vector spaces: quantum mechanics uses both finite and infinite-dimensional (Hilbert) space to represent states of a quantum system such as an electron. In addition, the theory uses operators on such spaces to represent various observable quantities of those states such as spin or momentum. The possible values of these observables are given by so-called eigenvectors, special vectors within the space that are invariant when the operators are applied to them. The various possible outcomes of an experiment have respective probabilities attached to them according to the size of the components of the eigenvectors.

The picture of fundamental particles that emerges from the mathematics is fundamentally mysterious. A photon traveling from a source to a detector exists as a wave, which amounts to a probability distribution. Where will it appear on the phosphor screen? Who knows—until it appears? Schrödinger's major contribution to quantum mechanics, the equation that now bears his name, describes the probability distribution of an electron in an atom. The following figure shows such a distribution for a hydrogen atom in an excited state. It looks somewhat like a pumpkin. If one measures the actual position of the electron, it might turn out to be anywhere within the pumpkin with equal probability. However, it will not appear very near either pole of the pumpkin, since the distribution does not allow it.

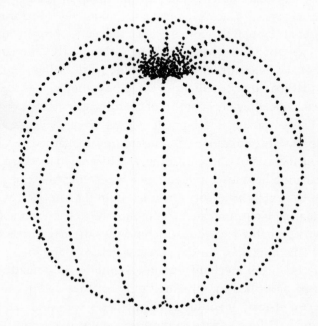

An excited hydrogen atom

Since the position of the electron is unknown until it is measured, one might even say that it has no real existence until that happens. But everything in the universe (more or less) is made from atoms the properties of which depend in the most fundamental way on the shapes of these probability distributions. If we run headfirst into a stone wall composed entirely of unreal atoms, we nevertheless end by being persuaded of its reality.

THE COPENHAGEN INTERPRETATION

The central question surrounding the dynamic attributes of a fundamental particle is whether these attributes may be said to possess definite values when they're not being measured. The question reminds us of the old chestnut "If a tree falls in the forest and no one is there, does it make a sound?"

Bohr's interpretation of the quantum facts and of quantum theory led him to believe that before a dynamic attribute of an electron, photon, or any other particle/wave was measured, the attribute had no particular value. For example, until a photon manifested itself as a point of light, it had no particular position. Until its polarization was measured, it had no particular spin. This interpretation of quantum mechanics is called the "Copenhagen interpretation," after Bohr's hometown. Bohr and, ultimately, the vast majority of physicists took the view that there was simply no underlying reality. Until the polarization of a photon was measured, it had no polarization at all.

The ultimately mysterious event during which a wave becomes a point of light on a screen, or takes one slit rather than another, is sometimes called "the collapse of the wave function." A wave function would appear to carry all of its potential values-as-measured simultaneously— pregnant with possibilities, so to speak. The collapse of the wave function is essentially a birth event, the ultimate process by which our classical "lived world" becomes manifest.

Einstein, among others, bridled at the mere thought of such indeterminacy and sought, by means of a series of thought experiments that he proposed to Bohr over a period of twenty years, to upset the quantum mechanical applecart by imagining a situation where quantum mechanics would either contradict itself or, at best, be helpless to provide an answer.

Einstein believed that our ignorance of the state of a particle/wave was "classical" in nature: we simply did not know enough about it. Bohr believed that our ignorance was "quantum mechanical" in nature, and we can *never* know more about it.

For every thought experiment that Einstein sent to Bohr, the latter always found a mistake or hole in Einstein's reasoning. Finally Einstein teamed up with two other physicists—Boris Podolsky, a Russian physicist who immigrated to America, and Nathan Rosen, an American—to propose the now famous EPR (Einstein-Podolsky-Rosen) thought experiment in 1935.

Originally proposed as an experiment involving the momentum of particles, an updated version uses polarization of photons. There are certain atoms, such as mercury, that when excited, emit not one but two photons when dropping back to their ground state. The two photons are said to be "entangled," owing to a definite but mysterious relationship between them. No matter where or when their polarizations are measured, they always turn out to be the same. According to the Copenhagen interpretation, neither photon actually had any polarization when it left the excited mercury vapor, yet somehow, when either was measured, it would always yield the same result as the other photon. The two photons were linked by a powerful relationship that transcended space itself.

Every fundamental particle possesses a quality called spin. Quantum mechanics aside, it does no harm to think of the particle spinning about its axis, with the spin axis pointing in a particular direction. The direction of this axis is called the polarization of the particle. When the spin of a photon is measured, a definite result is always obtained. Because only one direction can be measured at a time, the outcome of such experiments always has a yes/no quality.

To measure polarization, physicists use a crystal of calcite, commonly called Iceland spar. A beam of unpolarized light passing through such a crystal splits into two beams that are refracted at different angles. The polarizations of the two beams are at right angles to each other.

If we now imagine a stream of photons entering a calcite crystal, some will take one path, the rest taking the other. If you imagine a photon with a particular polarization entering the crystal, you might be puzzled by the fact that even if its supposed polarization is not the same as either of the exit paths, it simply will have the polarization attributed to the particular path it took.

For example, if you send a stream of random photons through a crystal with one exit stream polarized in a 45-degree diagonal direction, you may measure these photons with a second crystal having the same orientation and you will discover that all of them take the same exit path. However, if you now change the angle of the second crystal to a horizontal orientation, you will find that half the photons exit by one path and half by the other.

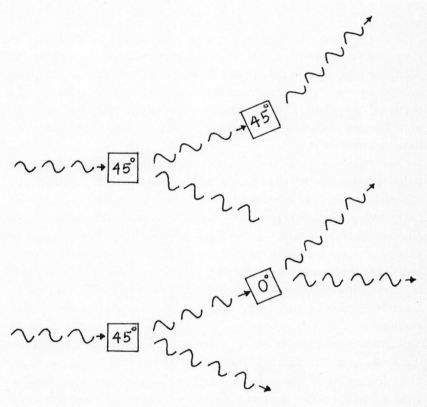

A stream of polarized photons

Which photon will take which path? No one knows. The same quantum indeterminacy that made it impossible to predict which slit an electron would pass through in the example at the beginning of this chapter also governs the behavior of photons.

The EPR thought experiment involves a source of entangled photon pairs, as described earlier. The experimental setup is shown in the figure.

The EPR experiment

From the source, S, a pair of photons is emitted as frequently (or infrequently) as one likes. The green photon travels to a detector nearby, while the blue photon travels to a detector on Mars, say. Here comes a photon now. The green photon detector, which happens to be set up at the angle α, goes off. Thus the green photon has a measured polarization of α. We now know that its blue twin, still on the way to Mars, also will have polarization α.

According to the Copenhagen interpretation, this photon has no polarization at all and won't have until it is measured. But how can it be, argued Einstein, Podolsky, and Rosen, that a photon can have no polarization when we already know how the measurement will turn out? After all, no action taken here on Earth can possibly affect the outcome on Mars. For such a thing to happen, the influence would have to travel faster than the speed of light, and nothing can travel faster than the speed of light. This property of experimental independence assumed by Einstein, Podolsky, and Rosen has come to be known as the locality assumption.

Einstein's chief aim was not to prove that quantum mechanics (the theory he helped to develop) was wrong, merely that it was an incomplete description of reality. Here, based on the simplest of assumptions, such as locality, was a result that begged for a different interpretation of events than the Copenhagen interpretation was willing to allow. Bohr answered the EPR thought experiment, called by some a "paradox," with the intellectual equivalent of a shrug. The attributes of a photon measured here or at a distance are the joint creation of the photon and the measuring apparatus, and that's all that can be said. As for this property of locality, who knows?

It is ironic yet strangely heroic that Einstein, who struggled to prove that there is more to physical reality than is described by standard quantum mechanics, should have produced a thought experiment that would ultimately undermine his own assumption of localized influences. From 1935 until the early 1960s, more than a hundred papers were written about the "EPR paradox." No resolution was reached until an Irish physicist, John Bell, took a sabbatical in 1964 and decided to have a long, hard look at the EPR thought experiment. Within the year, he had discovered a mathematical result that, when compared with the predictions of quantum theory, revealed a nonlocal universe, one in which the results of a measurement here could immediately and instantaneously affect the outcome of a measurement over there.

BELL'S THEOREM

I will explain Bell's theorem by first divorcing it entirely from its quantum mechanical context. Suppose I make up a table with three rows labeled A, B, and C. Each column of the table will consist of 0s and 1s, which I may write in completely arbitrary fashion. The following table will serve as an example; even the number of columns in the table is completely arbitrary:

A	0	1	0	1	0	0	0	1	1	0
B	1	1	0	1	0	0	1	0	1	1
C	1	1	1	0	0	1	0	0	1	1

For the sake of a verbal handle, let us call such an object a "table of triples." I will also introduce a function $XY(x, y)$, where X and Y will be two distinct row letters and x and y two entries under the same column in the respective rows. The function XY simply tells us the number of times that an x appears in row X in the same column that a y appears in row Y. In the example above,

$$AB(1, 1) = 3,$$

while

$$BC(0, 1) = 2$$

and

$$AC(1, 1) = 2.$$

We note in passing that, in this particular example,

$$AB(1, 1) + BC(0, 1) \geq AC(1, 1),$$

that is,

$$3 + 2 \geq 2.$$

It is now easy not only to state Bell's theorem but also to prove it:
Theorem: For any table of triples,

$$AB(1, 1) + BC(0, 1) \geq AC(1, 1).$$

Proof: The simplest proof of this inequality observes that each column of a table of triples, by itself, obeys Bell's inequality. If I simply write any triple of binary numbers, a b c, and assume that the inequality does not hold, then $AC(1, 1)$ must equal 1, and the sum of the terms on the left side of the inequality must be 0. But if $AC(1, 1) = 1$ it must be true that

$$a = 1 \quad \text{and} \quad c = 1.$$

What value can the binary number b have? If $b = 0$, then

$$BC(0, 1) = 1 \quad \text{while} \quad AB(1, 1) = 0$$

and the inequality holds. Since the inequality was supposed to fail, it must be true that $b = 1$. But this assumption also contradicts the supposition because in this case

$$BC(0, 1) = 0 \quad \text{while} \quad AB(1, 1) = 1.$$

I have just shown that every column of a table of triples obeys Bell's inequality. If I now add the contributions of every column to each of the three functions, I will always be adding function values that obey the inequality separately and, therefore, collectively as well. This proves Bell's theorem.

If the reader began to nod off during this proof, I can hardly blame him or her. I nearly fell asleep myself. To some people it may sound snooty to say so, but as mathematical theorems go, it is strictly ho-hum. Given a certain interpretation however, the theorem suddenly develops a profound significance—not mathematical, but physical. To explain the significance, we must revisit the EPR thought experiment, as Bell did en route to his theorem.

Suppose we measure the polarization of a single photon in three different directions. In each case the number in parentheses will be used as a code to symbolize the outcome of the measurement in question:

A: parallel (1) or perpendicular (0) to the horizontal

B: parallel (1) or perpendicular (0) to an angle α, and

C: parallel (1) or perpendicular (0) to another angle β.

Unfortunately, we cannot make all of these measurements for a single photon because the act of measurement of polarization requires that it pass through a polarizing crystal, which spoils all subsequent measurements. However—conceptually, at least—we can imagine measuring its polarization in the horizontal direction, then turning back the clock and remeasuring the same photon's polarization in the direction α, then turning back the clock once more to measure the β-polarization.

If we were to repeat this conceptual experiment thousands of times, we would end with a table of triples to which we could then apply Bell's theorem, finding that the inequality is satisfied, as it must be.

A more practical way to produce a table is to send out photon pairs, as in the EPR thought experiment, from a central source to two other points, called "here" and "there." If we measure the horizontal polarization of one photon here and measure the α-polarization there, we would get a pair of numbers that reflected the results of those measurements. If the numbers were 0 and 1, for example, the photon would not be polarized in the horizontal direction but would be in the α-direction. This single little experiment would contribute 0 to the term $AB(1, 1)$. If the photon and its twin both produced 1s, however, the contribution would be 1. So let us make such a measurement on one thousand pairs of photons, obtaining, in the process, one thousand values contributing to the quantity $AB(1, 1)$, some of them 0s and some of them 1s.

We could certainly construct a partial table with the first two rows filled in. We could even compute $AB(1, 1)$. Indeed, if we carried out this extensive experiment many times, the values of $AB(1, 1)$ that resulted would all be rather similar, subject only to slight statistical fluctuations.

If we now perform the second experiment, measuring polarization at the angle α over there, repeating it thousands of times as well, we could get a second partial table with the second and third entries in each row filled in. If the property of locality holds, the results of this experiment should be completely independent of the first experiment, at least in a statistical sense. The value of BC(0, 1) should in no way depend on the value of AB(1, 1). In fact, we should then be able to treat both sets of entries as if they came from the same table, fully expecting them to satisfy Bell's inequality. The same thing is true for AC(1, 1) when we finally compute it after our lengthy experiment.

Assuming locality, we may then infer that Bell's inequality holds for these observations and that when the experiments are all complete and the numbers toted up, we get

$$AB(1, 1) + BC(0, 1) \geq AC(1, 1).$$

When the theoretical predictions of quantum mechanics are compared with Bell's inequality, there are angles α and β, which produce predicted values that contradict Bell's inequality. For example, with $\alpha = 20$ degrees and $\beta = 60$ degrees, the actual values for these functions, as predicted by quantum theory and as verified by experiment, are

$$AB(1, 1) + BC(0, 1) = 0.66$$

and

$$AC(1, 1) = 0.88.$$

Has quantum mechanics failed? Since Bell's inequality holds for all possible tables of triples, it certainly didn't fail. The fact that the two outcomes are different lies with the fact that the three numbers were computed on the basis of three separate tables that, it must be concluded, were not equivalent at all. They were not statistically independent because, somehow, the measurements that produced the AB count were strongly influenced by the measurements of the BC and AC counts.

It would be astounding to most physicists if, after nearly a century of rigorous testing, quantum mechanics should suffer such a failure. It is by far and away the most successful physical theory we have ever discovered. On the other hand, many physicists would be upset by the idea of nonlocality. How to decide between the alternatives?

In a purely mathematical setting, the situation would be analogous to Gödel's theorem: either a system containing the standard arithmetic is inconsistent, or it contains theorems that cannot be proved within the system. (See chapter 6.) It is not clear how mathematics could decide which of the two alternatives held for a given system, but physics has the advantage of experiments. Is there an experiment that would demonstrate a clear violation of Bell's inequality and thus directly imply that quantum mechanics is a nonlocal theory?

Bell's theorem first appeared in a little-known journal in 1964. It took several years for most physicists even to become aware of Bell's theorem. It fell to a newly minted Ph.D., John Clauser, to actually test quantum mechanics against Bell's theorem in 1972 at the University of California at Berkeley. Clauser apparently thought he might disprove quantum mechanics by demonstrating that the predicted violation of Bell's inequality simply did not happen. He used excited mercury atoms as sources of twin photons, changing the orientation of his polarization detectors 100 times a second. His detector data, when analyzed, revealed a clear violation of Bell's inequality, just as quantum theory predicted. After that, it became increasingly difficult for physicists to ignore Bell's inequality and its violation by actual quantum mechanical experiments. Nevertheless, some physicists argued that the detectors, being in the same room, might still be interacting by an unknown yet strictly local mechanism.

French physicist Alain Aspect removed even these objections by carrying out essentially the same experiment but with a detector-switching time of one ten-billionth of a second. This meant that no local interaction could take place between the two meters because the detectors changed their orientation while each photon pair was in flight, so to speak. The switching time was faster than it took a photon to cross the laboratory. By 1982, Aspect announced his results. The world was definitely nonlocal. The effects of nonlocality extend well beyond the reach of laboratory apparatus. Every fundamental particle in our bodies has interacted with untold billions of fundamental particles everywhere else in the universe, and all are to some degree entangled with each other. There is an unseen but possibly influential synchronicity in our affairs.

THE MEASUREMENT PROBLEM

Up to this point I have pretended that quantum theory is a monolithic scientific enterprise in which all physicists are in agreement. While there is little disagreement about the mathematical and operational side of quantum theory, there are several schools of thought about the underlying reality.

The best introduction to these schools takes the so-called measurement problem as its starting point. This problem centers on the question "When does the wave function collapse occur?"

The question sits exactly on the boundary between the quantum world and the "classical" world. Essentially, the quantum world involves very small objects, while the classical world consists of all the objects we can see, feel, or sense directly.

Hungarian-born John von Neumann, perhaps the greatest mathematician of the twentieth century, published a classic book on quantum mechanics in 1932. Called *Die Grundlagen,* von Neumann's book painted an all-quantum picture of the world. Even macroscopic objects such as your body or the Earth itself had its own proxy wave with its associated quantum numbers. On the one hand, von Neumann demonstrated that electrons and other fundamental particles are not "real" entities because they cannot be said to possess any dynamical attribute before it is measured. This seemed to buttress the philosophical approach of the Copenhagen interpretation. On the other hand, von Neumann did not endow measuring instruments with a special status, as the Copenhagen interpretation did.

Von Neumann analyzed the measurement act, breaking it down into a series of steps called the "von Neumann chain." Applied to the two-slit experiment, for example, the von Neumann chain consists of (1) the emergence of a photon from a source, (2) its passage through one of the slits, (3) the triggering of a detector, (4) the signal from the detector to a meter, (5) the movement of a needle or other registration device, (6) the light from the meter to the eye of the observer, (7) the message from the observer's retina to the observer's brain, (8) processing of the signal in the observer's brain, and (9) registration in the observer's consciousness.

Where does the collapse occur? The problem of locating the collapse is well illustrated by the story of Schrödinger's cat. A live cat, along with a photon source, a pair of slits, a detector, and a loaded revolver are placed in a sealed, soundproof, lightproof box. Inside the box, things are arranged so that if the photon passes through one slit, the revolver is triggered and the cat is killed. But if the photon passes through the other slit, the revolver is not triggered and the cat remains alive. According to the Copenhagen interpretation of quantum mechanics, the cat is neither dead nor alive until the box is opened. Is this a reasonable proposition?

Von Neumann showed that one may place the collapse at any point on the chain that one likes. Only one site, however, has anything like a privileged position in the chain: the consciousness of the observer's mind.

Many people have drawn silly conclusions from this hint that consciousness may play a special role in the collapse of the wave function. For example, one school of thought has concluded that human consciousness literally creates the world. It may be that any attempt to peer behind the quantum curtain results in silliness of one kind or another. Why should I, in trying to discern some shape behind it, not be silly as well? Could it be that our own consciousness results from an ongoing process of collapse in our own brains? If so, strange as it may sound, the reverse also may be true: wave function collapse is the result of a conscious process—not necessarily our own, however.

What *is* going on behind the quantum curtain?

IS THERE A WAY AROUND IT?

One of the major non-classical features of quantum mechanics is the randomness associated with the collapse of wave to particle and its intimate connection with the measurement problem. While still a graduate student at Princeton in 1957, Hugh Everett III proposed that each time a particular outcome emerges from a quantum event, the universe splits into as many copies as there are outcomes. Each copy is identical except for the outcome of the measurement. In one of these universes, you and I see a point of light *here* on the screen but, simultaneously, copies of you and me in another universe see the point of light *there*. Since quantum collapse is going on all the time, even apparently when unwitnessed by us, our universe spawns myriad others at every moment. Each universe follows the same laws of physics as our own but remains forever separate. This proposal is called the "many-worlds hypothesis."

The many-worlds hypothesis certainly solves the measurement problem; all possible outcomes actually do occur, each in its own universe. To many, however, the proposal seems extravagant beyond measure, a cure that is worse than the disease. The many-worlds hypothesis is nevertheless regarded as serious a contender as the preferred explanation of the collapse of the wave function, a fact that signals both the frustration of physicists and the impenetrable nature of the quantum curtain.

Another way around the quantum collapse problem involves an entirely new interpretation of quantum theory, worked out in 1951 by David Bohm, an American expatriate physicist living in England. In Bohm's theory the wave function is a real but undetectable wave called the pilot wave. As the name suggests, it shepherds a single particle through slits, polarizers, and magnets, playing the wave role as witnessed in the laboratory, while the particle, equally deterministic, passes through one slit or another, depending on its actual position, also unmeasurable. Bohm's theory could be called an all-classical version of quantum mechanics. Unfortunately, the theory requires that the pilot wave communicate information about the environment instantly to its associated particle, invoking superluminal signaling velocities in which real information is communicated.

Some have argued that the entanglement phenomenon ought to make superluminal messaging possible. The idea is that since both experimental parties (green detector and blue detector) have the same information (the polarization of a photon pair), communication between them should be possible. But here again quantum mechanics draws its veil across our knowledge. Randomness.

The only way we can transmit a message is to have the ability to manipulate the polarization of individual photons, in sequence, as they are produced. According to the principles of quantum mechanics, this is impossible. No one can predict, much less manipulate, which polarizations will emerge when an atom such as mercury begins to give off photons. A random message has zero information.

On the other hand, the twin-polarization phenomenon promises a breakthrough in secure communication. When two parties wish to exchange secret messages, they generally do so by encrypting the information in a code. Virtually all codes in use today are enciphered and deciphered using a key specially set up for the purpose. Therefore there is always danger of the key falling into the wrong hands. For many kinds of encryption schemes the key can be an arbitrary string of zeros and ones. What better than a stream of randomly polarized photons? Both parties, separated by as great a distance as you like, can simultaneously receive precisely the same key and thereafter send each other secure messages. Moreover, should one of the photon streams such as the blue one be intercepted, this would amount to a measurement act, destroying the key and rendering it useless. The interception would be immediately observable at the blue end of the channel. For example, if the interloper measured polarizations with a calcite crystal, the blue end might receive photons that all had the same polarization.

The world of quantum phenomena is markedly different from our own. No doubt, stranger discoveries await us as we drift toward a view of the ultimate reality first hinted at by Einstein when he said, "Not only is the universe stranger than we think, it is stranger than we *can* think."

The Edge of Chaos

Unpredictable Systems

THERE ARE SOME CLASSICAL SYSTEMS (SUCH
AS THE WEATHER OR PLANETARY MOTIONS)
THE LONG-TERM BEHAVIOR OF WHICH CANNOT
BE PREDICTED BY ANY MATHEMATICAL LAW
(OR COMPUTER).

THERE CAN BE LITTLE DOUBT that the term "chaos" is far more exciting than the term "extreme sensitivity to initial conditions," but it is
nevertheless somewhat misleading. The one-word label has come to be
attached to a wide variety of phenomena that are predictable in principle but not in practice. The phenomena all reside in what physicists call
dynamical systems, arrangements of physical objects that, once set in
motion, follow rigid laws that should make the behavior of the system
completely predictable. Where's the problem?

The behavior of such systems may depend critically (and not just approximately) on their current states. A tiny change in a current state can completely alter future states in a surprisingly short time. And the "tiny change" may be too small for a computer to register.

My favorite example of a chaotic dynamical system is that old-fashioned taffy-mixing machine, the one with two rotating ladle bars that stretch and fold the soft delicacy over and over again, once every 2 seconds or so. When a portion of the taffy is stretched, two particles of the material (let us call them taffy molecules) draw somewhat apart. When the metal arm swings by again, one of the particles may be caught up in it more than a neighboring particle is. Before you know it, the former neighbors are at opposite ends of the mixing tub. The history of a particle depends critically on its initial position. That is chaos.

In terms of the way language is ordinarily used, there's nothing "chaotic" about the mixing process. Indeed, it's quite orderly. Replace the taffy by a mathematically idealized, viscous fluid, write the equations that represent the mixing process, and you couldn't ask for a more deterministic system. Yet we cannot always predict where a given particle will end up, especially after several cycles of mixing.

The problem lies not only in the process but also in the way we represent it. Suppose I have coordinates in centimeters for the position of a particle of idealized taffy at a particular moment:

(21.2732, 51.3725, 0.8226).

Where will the particle be exactly 2 seconds later? Surely, with that many digits of accuracy, I ought to be able to predict its position. I will use a computer microscope.

I also will examine a second particle, this one a near neighbor, under the microscope:

(21.2733, 51.3725, 0.8226).

This particle is a mere micron (one-thousandth of a millimeter) away from the first one in the direction of the first coordinate. The question now becomes, "Where will the two particles be 2 seconds later? As the virtual taffy undergoes the throes of equational mixing in the computer, we may find that after 2 seconds the two particles are 3 microns apart. Then along comes the mathematical mixing bar, stretching the conceptual fluid so that, 2 seconds later still, the particles are now 5 microns apart. This isn't too surprising when you consider that the taffy is being pulled every time the mixing bar comes around. But suppose now that the first particle of taffy happens to be rather close to the mixing bar, while the second is farther away, so that when the taffy is pulled, the second particle ends up nearly 100 microns from its previous neighbor. On the next go-round, the particles could be separated by as much as a millimeter, then a whole centimeter. After surprisingly few revolutions of the bar, the two particles could be on opposite sides of the mixing tub.

So where is the "chaos"?

In case you wondered why I bothered specifying all those digits, the grim truth can now be revealed. If I had specified only three digits after the decimal point instead of four, the two particles would have been indistinguishable as far as their initial positions were concerned. How then could a computer registering only this many digits possibly have predicted where *the* particle at that position might have ended up several revolutions later, when a difference of one unit in the fourth position could make such a huge difference? The short answer is that it could not have made such a prediction with any expectation of accuracy.

In the conceptual taffy, there is no limit to how many decimal digits I might need to predict where a given particle might be at a later point in the machine's operation. Indeed, only if I am allowed the power of infinite decimal expansions to express the positions of points in the virtual taffy will the mixing equations make accurate predictions. Computers, of course, are not designed to use infinite decimal expansions, nor could they be.

The crucial point at which the particles first begin to separate dramatically typifies what we call "extreme sensitivity to initial conditions," or "chaos" for short. In most dynamical systems capable of chaotic behavior, it does not happen all the time, only sometimes. But when it happens, all predictions go out the window. The phenomenon is sometimes also called the "butterfly effect," a metaphor that expresses the presence of chaos in weather systems. Thus, a butterfly beating its wings somewhere in the Amazon rain forest today will make all the difference in Holland a week later between a heavy windstorm and a nice day.

It is highly doubtful that a butterfly has ever had this effect in the history of the planet—but who knows? As we will see, weather systems appear to be chaotic in this sense, so it is possible that a zephyr breeze developing along the coast of India, had it been just a trifle weaker, would not have contributed to a nasty typhoon in Indonesia four days later.

In the mixing process just described, not only do particles separate, but they come together as well. Chaotic systems have been characterized by the American mathematician Stephen Smale as consisting of two operations: stretching and folding.

THE "SCIENCE" OF CHAOS

Although it has been called a "science," it would be more accurate to describe chaos theory as a "field" within science. Research in this field consists of the discovery of new chaotic dynamical systems, as well as the development of general theories of structure and detection. As such, chaos theory may be located within physics, although like many areas within physics, it tends to be rather mathematical in nature. One major feature of this new field is the status of experiments. Most of the discoveries were made using not real systems, but computer simulations of them. The logical structure of the model in each case produced phenomena that researchers could assume also occurred in nature itself. The computer, meanwhile, plays the villain of the piece. If only it could handle numbers with an infinite number of digits!

Unlike many fields, chaos theory began in several widely scattered locations at equally scattered dates, ranging through the 1960s and 1970s. A handful of researchers in several different fields studied the new phenomenon of extreme sensitivity to initial conditions, few of them being aware of other work in the newly emerging field. In the next sections I will describe the work of just two of these researchers in the 1960s, Edward Lorenz at the Massachusetts Institute of Technology and Robert May at the Princeton Institute for Advanced Study. Lorenz studied models of weather systems, while May agonized over the behavior of a seemingly simple predator-prey model. I will begin, however, with a third player.

In what can only be described as the blossoming of a zeitgeist, the 1960s saw the first great surge of exploration of chaotic systems. As well as Lorenz and May, mathematician Stephen Smale turned from the study of topological surfaces to a study of dynamical systems.

As we will see later, every dynamical system can be represented in a phase space where its behavior can be described as a trajectory. As the system goes through its motions, a point in phase space traces out a path that may be straight, curved, or even hopelessly convoluted. In the most general view, the trajectory either lies on or defines a shape in phase space, usually a surface of lower dimension than the phase space itself. Such shapes amounted to surfaces or "manifolds," as topologists call them, and thus it was not at all surprising that Smale would take up the study of these particular kinds of manifolds.

Smale was particularly concerned with the stability of dynamical systems, assuming that all or most of them would eventually settle into some equilibrium state. He had a hunch that the topology of his dynamical manifolds would reveal an answer. His hunch was right, but his assumption was wrong. The behavior of many dynamical systems could be mirrored in a model that featured a succession of operations in which the phase space was molded, so to speak. Stretching and folding (like the taffy machine) turned out to be a major clue to the behavior of such systems. Thus Smale first became aware of the extreme sensitivity to initial conditions that practically typified dynamical systems.

In the early 1970s, James Yorke, a mathematician at the University of Maryland, stumbled on a paper by Edward Lorenz that had appeared in an obscure journal. Yorke, with colleague T. Y. Li, studied the new phenomenon of sensitivity, especially fascinated by the boundary between chaotic and nonchaotic behavior in dynamical systems. The pair published a now-classic paper with the intriguing title "Period Three Implies Chaos." Yorke had perhaps intended the last word as a mild joke, but the name stuck. As it happened, the word "chaos" not only described Yorke's emotional reaction to the utter unpredictability of such systems but also would spin the new field into prominence in the media.

Benoit Mandelbrot, who had spent decades laboring in obscure areas of applied mathematics, from analyzing patterns in cotton prices to measuring coastlines, made his research home at the IBM Yorktown research center. Mandelbrot was interested in the appearance of scaling phenomena in data from a wide variety of sources. For example, price fluctuations could occur at all scales, from a few cents to many dollars. Moreover, many natural objects, from clouds to cauliflowers, presented the same phenomenon of scaling, showing the same features at different scales of magnification. Spaces with such a scaling feature often had no straightforward dimension. For example, what is the dimension of a ball of twine? At a distance, it appeared to be three-dimensional. Closer up it appeared to be one-dimensional, being nothing more than a linear element (the twine itself) wound up in a ball. Yet the twine had thickness, so surely it was three-dimensional. But wait. The twine was composed of one-dimensional fibers! In the end it made sense to assign a fractional dimension to such objects. It was Mandelbrot who coined the term "fractal," meaning an object with fractional dimension.

In the early 1970s, Harry Swinney, a physicist at the City College of New York, collaborated with colleague Jerry Golub to study phase transitions in fluids. The pair used a simple apparatus in which fluid was set into motion between two cylinders. As the outer cylinder, made of glass, was rotated faster and faster, the fluid between the cylinders showed transitions from smooth to turbulent flow. At specific speeds there would be a transition from turbulence with one frequency of eddies to turbulence with a higher frequency. Then at an equally specific point, the turbulence became utterly disorganized—chaotic, one might say.

Meanwhile, a French physicist, David Ruelle at the Institut des Hautes Études Scientifiques in Paris, along with a Dutch colleague, Floris Takens, made an amazing discovery about such disorganized forms of turbulence. They had found rather special structures in the phase space of such systems. They called the structures "strange attractors." Strange attractors began to show up everywhere, it seemed, from systems of electronic oscillators to the motions of planets in the solar system. Chaos was ubiquitous.

THE LOGISTIC EQUATION

If there's one accessible dynamical system that captures the essence of chaos in a simple but precise manner, it's the now-famous logistic equation. The logistic equation is simply the reincarnation of an equation discovered by the nineteenth-century Belgian mathematician François Verhulst, the first to use it to explore the ups and downs of populations.

It was also in the nineteenth century that Malthus had pointed out that populations of organisms, if supplied with unlimited resources, would enjoy exponential growth in their populations, increasing their numbers by a fixed ratio r over a period of time, then increasing them again, and so on without limit. Thus, if the size of a population growing according to Malthus' prescription were x at the beginning of the period of time in question, it would be rx at the end of the period, where r is the ratio of increase.

The prospect of Malthusian growth influenced the development of the Darwin-Wallace theory of evolution. Owing both to competition and to limited resources, most populations do not grow in this fashion except rarely or periodically. In the twentieth century a few theoretical biologists played with the Verhulst equation in the hope that some rays of light might be thrown on how real populations behave. In the Verhulst equation, the population at the end of a period of time would depend not only on x, but also on the level a of available resources that, when consumed by the x organisms, would be depleted to a level $a - rx$, in effect.

The puzzling thing about the Verhulst equation was its behavior. Sometimes it gave very nice, well-regulated populations, but sometimes the numbers would vary all over the map in the most confusing way.

In the 1960s, at the Princeton Institute for Advanced Study, Robert May, a physicist turned theoretical ecologist, began to explore the Verhulst equation, determined to get to the bottom of the equation's strange behavior. May simplified the equation, renaming it the "logistic equation," meaning that growth of a population would be limited by the logistics of the environment in which it found itself.

The logistic equation describes an abstract relationship between the relative abundance x of a predator and the relative abundance $(1 - x)$ of its prey. For the predator I will use a made-up animal, since then I don't have to worry about being unrealistic. The animal will be called a grim-blik, and currently in our pretend ecosystem there are 438 grimbliks. On the other hand, there are 10,255 smorts, the favorite prey item of grimbliks. Now the total biomass of both creatures we will take as $438 + 10,255 = 10,693$ units, grimbliks and smorts having roughly equal weight. If I divide both abundances by the sum 10,693, the total biomass becomes 1 and the abundances become relative abundances of 0.04096 and 0.95904, respectively (to five decimal places, anyway).

In this conceptual system, the relative abundance x of new grimbliks after a certain period of time will be proportional to the product of the relative abundances of the two populations:

$$x(1 - x).$$

After all, the number of smorts eaten will depend not only on the number of grimbliks but on the number of smorts that happen to be around, as well.

By introducing a constant of proportionality, normally represented by the Greek letter lambda (λ), we may write the equation as follows:

$$x = \lambda x(1 - x).$$

The x on the right-hand side of the equation represents the former number of grimbliks, and the x on the left represents the new population level after the period of time has passed. This particular dynamical system, unlike the taffy machine, operates in discrete jumps, the intervening predation being assumed to be continuous. The constant lambda represents the "fecundity" of the population, or its tendency to increase. In other words, a low value of lambda implies a population that does not increase very much over the period in question, while higher values mean higher rates of increase.

How large could such values get? Since x must always be less than 1, being a proportion, the new population, x, would have to be less than 1. In others words,

$$\lambda x(1 - x) < 1.$$

It is a simple mathematical fact that the expression $x(1 - x)$ reaches its maximum value over the domain from 0 to 1 when x and $(1 - x)$ both equal $\frac{1}{2}$. The maximum value must therefore be $\frac{1}{4}$. The left-hand side of the equation above cannot exceed the value

$$\lambda(\tfrac{1}{4})$$

and, since this quantity must be less than 1, we conclude that lambda can have any value less than 4. In other words, whatever the fecundity of a population might be, it cannot exceed 4. Populations with low fecundity are very well behaved, according to the logistic equation. For example, if $\lambda = 2$, the Verhulst equation becomes

$$x = 2x(1 - x).$$

Suppose then that grimbliks have precisely this fecundity and that an initial population level (proportion) happens to be 0.8. In this case, the equation produces a succession of values for x when the equational process is iterated over and over again:

$$0.800 \rightarrow 0.320 \rightarrow 0.435 \rightarrow 0.492 \rightarrow 0.499 \rightarrow 0.500 \rightarrow$$
$$0.500 \rightarrow \cdots$$

In this calculation, I carried three decimal digits, and the values for x quickly converged to a number, namely 0.5. It would make no difference how many digits I carried in this calculation; the result would always be the same, ultimately 0.50000 $\cdots$ to as many digits as I like. There is no question of chaos in this toy predator-prey system. Ultimately the two populations settle down to the same values, 50 percent grimbliks and 50 percent smorts. Interestingly enough, it makes no difference what starting value I might use for x. The ultimate population levels would reach equality, at 0.5 each. In the language of dynamics, the system has one stable point, when $\lambda = 2$.

What happens if we make the grimbliks more fertile, setting $\lambda = 3$? Let's do the calculation and find out, starting with the same initial value for x. It takes a little longer to see what the system will do in this case, as its behavior is a bit more complicated.

$$0.800 \rightarrow 0.480 \rightarrow 0.749 \rightarrow 0.564 \rightarrow 0.738 \rightarrow 0.580 \rightarrow$$
$$0.731 \rightarrow 0.590 \rightarrow 0.726 \rightarrow 0.597 \rightarrow 0.722 \rightarrow 0.602 \rightarrow$$
$$0.719 \rightarrow 0.606 \rightarrow 0.716 \rightarrow 0.610 \rightarrow 0.714 \rightarrow 0.613 \rightarrow$$
$$0.712 \rightarrow 0.615 \rightarrow 0.710 \rightarrow 0.618 \rightarrow 0.708 \rightarrow 0.620 \rightarrow$$
$$0.707 \rightarrow 0.621 \rightarrow 0.706 \rightarrow 0.623 \rightarrow 0.705 \rightarrow 0.624 \rightarrow$$
$$0.704 \rightarrow 0.625 \rightarrow 0.703 \rightarrow 0.626 \rightarrow 0.702 \rightarrow 0.628 \rightarrow$$
$$0.701 \rightarrow 0.629 \rightarrow 0.700 \rightarrow 0.630 \rightarrow 0.699 \rightarrow 0.631 \rightarrow$$
$$0.699 \rightarrow 0.631 \rightarrow 0.699 \rightarrow 0.631 \rightarrow \cdots$$

Can you sort out what's happening here, amid the welter of numbers? The sequence of values for x produced by this equation ultimately turns out simply to alternate between two values, namely 0.699 and 0.631.

I have illustrated the behavior of May's system in a one-dimensional phase space. Each successive value of x is represented by a point. The convergence of the system to the two attractor points makes the overall process clear.

The grimblik population x oscillates and, of course, so does that of the smorts, since the relative number of smorts is always $1 - x$. Once again, it doesn't matter what initial population size of grimbliks you start out with, the end result is always the same. In successive generations, the population levels of grimbliks and smorts simply swap values. If smorts happen to have 69.9 percent of the total population in one generation, they will have 63.1 percent of the population in the next generation.

When λ has the value 3, in other words, the system develops what dynamicists call two stable points, alternating between the two values.

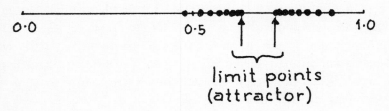

Phase space for grimbliks

May was well aware of this fact. The problems began when he carried out the calculations for higher and higher values of lambda. With a value of 3.5 for lambda, for example, the system developed four stable points, with the population of grimbliks jumping from one value to another, regularly visiting all four values and always in the same order. At the value of 3.56, May discovered that the system had suddenly developed eight stable points and just beyond this, 16, then 32, then 64, 128, 256, and so on. May knew perfectly well that something strange was in the offing; although lambda must always be finite—less than 4, in fact— the period of the system was doubling, with no end in sight. What happens when the infinite collides with the finite?

At the value of $\lambda = 3.569946$, all hell broke loose. There were no stable points at all! The population of grimbliks, in other words, showed no regularity whatsoever, hopping erratically from one value to another with no evident rhyme or reason.

May made a map of the behavior of his logistic system. For many values of lambda between 0 to 4 he stacked up the resulting phase spaces, plotting the points that the corresponding logistic systems converged to. It was sketchy but clear enough to show something extraordinary. The (single) stable points drifted to higher values for increasing values of lambda, forming a gentle curve that suddenly split into two curves, then four, and so on. The following figure shows a more precise version of the map, as produced by a modern computer.

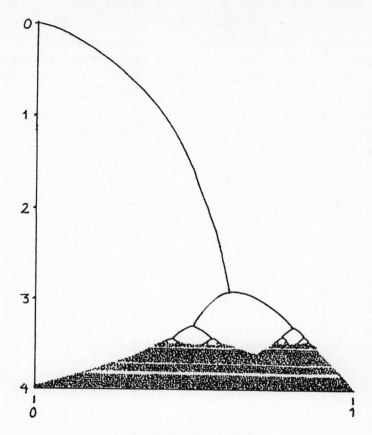

A map of the logistic system

Above the magic value of 3.569946 · · · , this map reveals what May could have seen only murkily—the onset of chaos. There are no stable points at higher values of lambda, except intermittently. Stability returns to the system, as indicated by the clear spaces with a finite number of lines, only to be wiped out once more by chaotic behavior.

What does the chaos in the logistic map have to do with extreme sensitivity to initial conditions? As we have already seen, for stable values of lambda you can start with any value of x that you like, and the system converges, or is attracted, to one or more stable points. The system is quite insensitive to initial conditions (the value of x) in such cases. On the other hand, the behavior of the system is quite unpredictable, in any practical sense, when the system has no stable points. It makes an enormous difference, in other words, how many digits your computer carries when portraying the behavior of the system. Let's try an experiment.

With a value of 3.6 for lambda, let's start with a precise value for x, say $x = 0.900$. I will iterate the logistic equation with this initial value, as well as with one that is very close to it, namely 0.901, until the two sets of numbers diverge substantially. This will mean that the two sets of values have drifted far enough apart for them to differ in the first digit to the right of the decimal point.

ITERATION NUMBER	FIRST INITIAL VALUE	SECOND INITIAL VALUE
0	0.900	0.901
1	0.324	0.321
2	0.788	0.785
3	0.601	0.608
4	0.863	0.858
5	0.426	0.439
6	0.880	0.887
7	0.300	0.361
8	0.756	0.830

After just eight iterations, the values of x have gotten far enough apart to differ in their very first decimal digit. But perhaps we weren't carrying enough digits. Let's try the same experiment with initial values of 0.9000 and 0.9001.

ITERATION NUMBER	FIRST INITIAL VALUE	SECOND INITIAL VALUE
0	0.9000	0.9001
1	0.3240	0.3237
2	0.7885	0.7881
3	0.6003	0.6012
4	0.8638	0.8631
5	0.4235	0.4254
6	0.8789	0.8600
7	0.3832	0.4334

Here, owing to the somewhat unruly nature of the logistic equation, the new sequences take even fewer iterations to become different in the first decimal place. It is irrelevant how many decimal digits my computer is able to handle. Sooner rather than later, its predictions about the number of grimbliks will be wildly wrong.

There is a structure—rather, a *kind* of structure—associated with all dynamical systems that are capable of chaotic behavior. That shape is what Mandelbrot called "fractal." The overall shape that I call a "dangling bell" occurs all over the diagram. Moreover, each little dangling bell contains an infinite regress of dangling bells! Notice that the fractal is not something you'd see during a walk in the woods. Instead, you have to visit phase space, as in the diagrams we examine in this chapter.

A fractal is any geometric shape that is composed, essentially, of copies of itself. In nature we see shapes that are suggestive of fractals: clouds and cauliflowers, trees and shorelines, to name a few. Although a cloud can be thought of as composed of clouds, and those clouds of cloudlets, the regress must come to an end quite abruptly and early. Water droplets are not shaped like clouds. Real fractals, on the other hand, just never quit! Within the dangling bell I have just shown you, other, smaller dangling bells await your inspection. All you need is a computer microscope.

STRANGE ATTRACTORS

It is no accident that fractals and chaos seem to occur together. First, every dynamical system has an attractor in its phase space, a set of behaviors to which the system is "attracted." Second, in the phase space of a dynamical system that is capable of chaos, the attractor always has a fractal shape. We will begin with a simple dynamical system and its phase space.

Here is a grandfather clock, with its steadily swinging pendulum. One phase space portrait of the pendulum's behavior uses the angle of the pendulum and its angular velocity as coordinates. For each point of time t in the progress of a dynamical system, these coordinates will have specific values. As time progresses, the system traces a curvilinear path through its phase space. For the pendulum in the grandfather clock, the

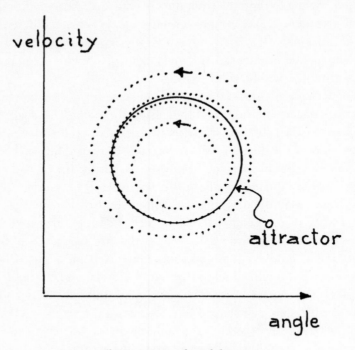

Phase portrait of pendulum

attractor is a (solid) circle, as shown in the phase diagram above. As the pendulum continues its stately swing, it continues to follow the circular path if it is not disturbed.

If we perturb the pendulum, either slowing it down or giving it a push, the escapement mechanism that drives the pendulum will bring it back to normal behavior, as shown by the dotted trajectories. Here, the track in phase space spirals back to the attractor. We witnessed similar behavior in the logistic system. At low values of the parameter λ, the population of grimbliks converged (or "was attracted") to the specific value of a one-point attractor. At higher values, the system converged to a two-point attractor, then a four-point attractor, and so on.

As far as the single pendulum is concerned, there is no chaotic behavior and no fractal in its phase diagram. But what about a double pendulum—that is, a pendulum with a freely swinging pendulum attached to its end?

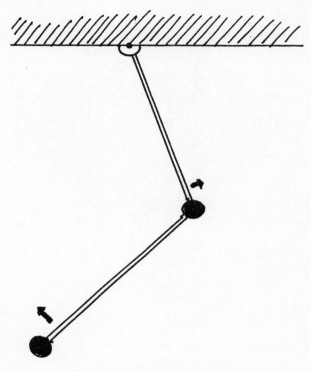

The double pendulum system

The motion of a double pendulum is very complicated. The upper arm of the pendulum does not swing with the straightforward back-and-forth motion of the single pendulum. Instead, it speeds up and slows down in its swing, sometimes completing only a portion before retracing its angles. If readers bestir themselves to do a search on the phrase "double pendulum" using their favorite Web browser, they will immediately find a host of engaging simulations of a double pendulum dynamical system. (Beware of hypnotic effects!) The double pendulum, once set in motion with the appropriate velocity, will display a crazy and unpredictable pattern of behavior, one that is, moreover, sensitive to initial conditions, one that is capable of chaos.

A phase diagram of the double pendulum system shows a complex fractal shape. The double pendulum is capable of chaos.

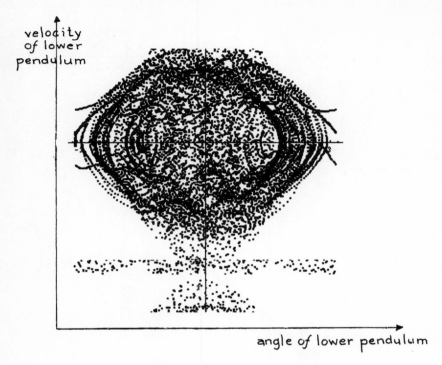

The double pendulum attractor

Although all chaotic dynamical systems have a fractal lurking in their phase portraits, not all fractals belong to dynamical systems. The mother of all fractals, the Mandelbrot set, is shown on the next page. Discovered mainly by Mandelbrot in 1981, the Mandelbrot set has been described as the "most complicated object in mathematics."

The Mandelbrot set may be generated by another iterative equation, namely,

$$z = z^2 + c.$$

This equation does not describe any dynamical system that I am aware of, although one might well be found. The issue is not particularly relevant because the Mandelbrot set is an object of study in its own right.

Without pretending to give a complete description of this equation, I will merely mention that the variable z takes complex values, namely numbers of the form $a + b$i, where a and b are real numbers and i is the

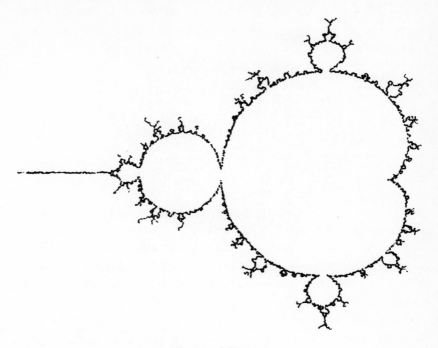

The Mandelbrot set

"imaginary number" that haunts the dreams of engineering students. Complex numbers exist not in a line, like real numbers, but in a plane called the "complex plane." Points in this plane have coordinates (a, b). When the equation is reiterated as the logistic equation was, the point z will jump from place to place in the complex plane, like a flea on a hot stove. The name of the game is always to start with the same value for z, namely $z = 0 + 0i$, the very origin of the plane. Of course, the value of the complex number c is fixed during the iteration process. If, during this process, the point z remains in the general area of the Mandelbrot set, it will do so forever and the constant c is then reckoned to be a member of the set. Otherwise, the point z will eventually flee to infinity, becoming larger and larger without limit.

When the constant c happens to lie in the Mandelbrot set, the behavior of the point z is often impossible to predict, owing once again to the same sensitivity to initial conditions that infects so many dynamical systems.

For example, if you start with a particular value for c, say $1.0 - 1.2i$, you may iterate the equation once to get from $z = 0 + 0i$ to $z = 1.0 - 1.2i$. On subsequent iterations z takes on a sequence of values:

$0.00 + 0.00i$

$1.00 - 1.20i$

$0.56 - 1.20i$

$-0.13 - 2.54i$

$-4.06 - 0.58i$.

The complex numbers in this progression, viewed as points in the plane, appear to be getting larger and larger without limit. And indeed they are. The original value of c, namely $1.0 - 1.2i$, must therefore lie outside the Mandelbrot set. However, I can give that point a color that depends on how fast it increases. Such colors provide the intriguing visual riots that characterize images of the Mandelbrot set.

To points within the set, we traditionally assign the color black. For example, the number $c = -0.5 + 0.1i$ produces the following behavior for z:

$0.00 + 0.00i$

$-0.50 + 0.10i$

$-0.26 - 0.00i$

$-0.43 + 0.10i$

$-0.33 + 0.02i$.

These numbers appear not to run "screaming off into infinity," as one of my math instructors used to say. If, indeed, the sequence of numbers remains bounded, then $c = -0.50 + 0.10i$ is indeed a member of the Mandelbrot set.

All about the Mandelbrot set one may see what some mathematicians call "mini-Mandelbrots" sprouting. And these mini-Mandelbrots have lesser mini-Mandelbrots and so on ad infinitum.

With this background we are now ready to examine the second featured scientist in our exploration of chaos.

THE LORENZ ATTRACTOR

In 1960 Edward Lorenz, a mathematician turned meteorologist, began working on the problem of weather prediction in a new way. He had noticed that, apart from seasonal variations and other periodic aspects, the weather in a given area never quite repeated itself. Were there patterns that nevertheless manifested themselves over time, patterns that weather forecasters might exploit?

Lorenz explored this question using a rather primitive desktop computer called a Royal McBee, a jumble of boxes, cables, and vacuum tubes he would program to simulate weather on a simple planetary surface that was represented by a two-dimensional grid of points. At each point he would specify all the relevant variables, such as temperature, air pressure, and humidity. He would also install all the relevant equations of fluid dynamics, heat transfer, and so on, hoping that in this toy system he might discover regularities or principles that, despite the model's simplicity, would also occur in the vastly more complicated ocean of air that enveloped the Earth.

Because computers of that era were far slower than today's machines and because the equations that Lorenz had programmed were extensive, he would initialize conditions at the points of his geographic grid and set the machine running. Then he would attend to other academic affairs, go for a walk, or visit the main office—unless the unwieldy machine had developed yet another bug. Lorenz found that his pretend planet also developed weather patterns that resembled those on Earth, at least insofar as they showed the same kind of variability.

Then, one fateful day in 1961, he decided to extend a run the machine had just completed. He could have simply started over, had the Royal McBee been a faster machine. Instead, he typed in the final grid values of the previous run at its midpoint, expecting that the machine would more or less duplicate its earlier performance before continuing the calculation beyond its original termination point. While Lorenz went for a coffee, the computer ground its way through the calculation, producing results that would startle him.

When he returned from his break, Lorenz found some puzzling numbers on the Royal McBee printout. The repeated portion of the calculation was nothing like the original!

After checking for bugs in his machine, he also checked the numbers he had entered for the repeat run. To save space on his printouts, Lorenz's program specified just three digits of output, instead of the six that the machine actually used. This was a perfectly reasonable thing to do since, to track the system's behavior, he did not need to see all six digits. If the air pressure at a particular grid point was actually 99.5327 in the machine, the value of 99.5 was just as informative as an indication of how the system was behaving.

To initiate the new run, Lorenz had simply typed in the approximate printout values, "knowing" that the repeated portion of the run would be little different from the original. After all, most systems of equations used in such scientific calculations were well behaved in this sense: a slight difference in input values would always produce a slight difference in output values. When he compared the numbers on the two printouts, however, Lorenz found that they diverged, at first a little, then a lot. How could this be?

Painstakingly, he analyzed the program's behavior on the data that had produced such a wide divergence in outputs. He isolated the equations that had been chiefly responsible for the difference, then created an even more miniature system that consisted of just one cell of the planetary surface. What he found both shocked and delighted him. Weather prediction would never be the same.

WEATHER IN A JAR

For the purpose of reproducing the anomalous behavior of his simulated weather system, Lorenz succeeded in reducing the number of equations to just three. The resulting minisystem consisted of a single air-filled cylinder. When heated, the air would rise in the center of the cylinder, descending along its sides.

As the air rose, it would cool at a rate determined by the value of a parameter that, like λ in the logistic equation earlier, could be set in advance. Another parameter controlled the amount of heat applied at the bottom of the column. Depending on the values he gave these parameters, his weather-in-a-jar system might or might not produce anomalous behavior.

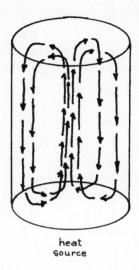

heat
source

Weather in a jar

For example, with one combination of parameter settings, Lorenz found that the column would begin to circulate, quickly settling down to a stable rate. But for other values, it would circulate for a while, gradually slowing, then reverse itself, the air rising along the sides and sinking in the middle, but then, later and in a seemingly unpredictable fashion, reverse itself once again.

Lorenz succeeded in boiling down the equations that were essential to the production of this unstable behavior to just three:

$$dx/dt = \sigma(y - x)$$

$$dy/dt = \rho x - y - xz$$

$$dz/dt = xy - \beta z.$$

In these equations, the variables x, y, and z each represent one aspect of weather in the jar, so to speak. The variable x is the velocity at which the column of air rises, while y represents the temperature difference between the ascending and descending air masses. The third variable is a little more tricky. In slowly moving air, the air temperature changes linearly as you go from one mass of air to the other. But as the air velocity ("wind") speeds up, this gradient becomes less and less linear, z measuring the extent of departure from linearity.

If you give the parameters the values σ = 10.0, ρ = 28.0, and β = 2.667, for example, you may solve these equations through the use of successive approximations, starting at $t = 0$, to obtain initial values for x, y, and z. For subsequent values of t, the differentials dx, dy, and dz give the increments in x, y, and z, enabling a computer to plot the behavior of the system through time.

The resulting picture, shown in the illustration, portrays the track taken by the system as the time, t, progresses. Readers must remember that the three-dimensional space playing host to this track is not the space inhabited by the jar. Rather, it is phase space, in which the coordinates represent air velocity, temperature difference, and temperature gradient nonlinearity. Can I simply look at this space and say, "Aha—I see it all very clearly"? No, I can't. But with some practice I can identify the right-hand part of the diagram as a place where air velocity is higher.

The resulting diagram somewhat resembles the butterfly in Lorenz's metaphor for chaos. The wings of this butterfly do not flap, however. Instead, they represent two sets of more or less circular tracks. The sys-

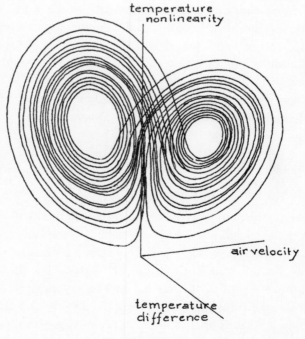

Lorenz's "butterfly"

tem will follow one set of tracks for a while, then, without warning, move into the second track, where the two sets intersect.

To develop our intuitions for the erratic behavior of the weather in a jar, we have the advantage of a very simple dynamical system that appears, at first sight, quite different from our little jar.

Imagine a waterwheel with buckets affixed to the outer rim. The buckets are filled by water that flows steadily from above the wheel. If the wheel has any motion, the buckets being filled naturally begin to pull the wheel increasingly in the same direction, the whole thing gradually speeding up. There is something peculiar about this particular waterwheel, however: the buckets all leak!

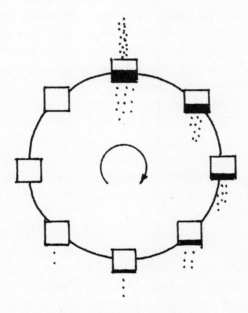

Lorenz's water wheel

If the wheel turns fast enough, each bucket spends less time under the waterfall as it passes beneath. It is even possible that the buckets currently being filled receive less water than the amount that remains in the buckets now ascending the left side of the wheel, for these were filled earlier, when the wheel was turning more slowly. Thus, if the buckets do not leak too quickly, the wheel may well slow down, stop, then reverse direction.

This peculiar waterwheel happens to mimic the weather-in-a-jar system exactly. The velocity of the wheel is simply the velocity of the air moving inside the jar. The leakiness of the buckets represents the loss of heat. If a column of air cools, it rises less quickly, just as the buckets on the descending side feel less attraction under gravity as they lose weight.

PREDICTING THE WEATHER

If the weather in a jar can behave so erratically, what about the real weather outside our windows? An argument for chaos in our weather could be sketched as follows: real weather is filled with columns of air— some rising, some falling. At the smallest scale, small parcels of air about the size of party balloons may rise from the heated ground. Over a patch of forest that is surrounded by fields, the air rises, producing the thermals enjoyed by birds of prey and glider pilots alike. In the fields around the forest, the air tends slowly to sink.

In a supercell thunderstorm, the frontal portion of the storm plays host to a vast column of rising air that may have a cross section of more than 100 square miles. Although not so nicely organized as the column of air in a jar, these columns, by virtue of containing the ingredients of Lorenz's chaotic system, will accordingly behave chaotically. A slight difference in such a column in the morning might produce a different sort of storm by late afternoon.

Even if our weather computers could handle data from every cubic meter of the Earth's atmosphere, they would be able to predict the weather days ahead with only occasional accuracy.

To see how modern weather forecasting systems are doing today, I performed a small experiment: I recorded the numbers from a great many four-day forecasts, noting the probability of precipitation (p.o.p.) and the predicted temperature (in degrees Celsius). I also noted the temperature and rainfall every day during the period, comparing them to the predictions of the previous four days. I continued the experiment for a little over 100 days, from early April until mid-July. That gave me enough data to provide a fairly reliable peek at success rates.

Predicted temperatures were all over the map, as can be seen in the accompanying table. Beside each possible difference between predicted and recorded temperatures, I have placed as many asterisks as there were instances of a prediction being off by that amount. Each distribution of differences centers roughly on a zero difference, and the average differences for the one-, two-, and four-day periods was within a degree of zero. The predictions showed a high degree of variance, however, with two possible manifestations of chaos in the one-day prediction period when the actual temperature exceeded the predicted temperature by 13 and 15 degrees, respectively.

HOW TEMPERATURE FORECASTS WORK OUT

	ONE-DAY	TWO-DAY	FOUR-DAY
−08	*		
−07	*	***	***
−06		***	****
−05	***	***	****
−04	*****	********	********
−03	*****	*******	******
−02	*****	**************	*******
−01	****************	***********	*****
00	***************	********	***************
+01	************************	************	**********
+02	***************	******************	*******
+03	**	***********	*******
+04	******	**	****
+05		*	**
+06	*		*****
+07			*
+08		*	**
+09			**
+10		*	**
+11			
+12			
+13	*		
+14			
+15	*		

Farther out in the prediction envelope, the guesses get wilder, even as they maintain an appropriate long-term average. This phenomenon can be seen in the shape of the distribution, the four-day predicted temperatures being distinctly more smeared out. Not surprisingly, the farther ahead we try to forecast temperature, the more likely we are to be wrong.

The analysis of the p.o.p. figures was even more startling. The accompanying table shows what each p.o.p. boiled down to in terms of what actually happened. Each percentage in the table proper represents the fraction of times it rained for a great many combinations of predicted p.o.p. figures and lead times (one-day, two-day, etc.). It must be allowed that many of the wilder outcomes represent samples too small to draw firm conclusions from. Taken collectively, however, one wouldn't expect quite so many wildly off figures.

WHAT "PROBABILITY OF PRECIPITATION" ACTUALLY MEANS

	One-Day %	Two-Day %	Three-Day %	Four-Day %	Average % (Number of Days)
00	8	22	0	50	19.4 (31)
10	27	0	0	20	15.8 (57)
20	21	33	38	43	31.6 (38)
30	20	27	36	14	25.3 (90)
40	20	46	41	46	41.6 (89)
50		0	0		00.0 (3)
60	86	50	57	73	65.0 (40)
70	33	50		0	33.3 (9)
80	90	86	86	80	86.2 (29)
90	100				100.0 (3)
100	66		0		57.1 (7)

For example, one would expect that if a 40 percent p.o.p. is predicted 100 times, it should rain on approximately 40 percent of those occasions. In fact, this is roughly what happened in the case of the 40 percent p.o.p.

prediction. This p.o.p. was predicted a total of 89 times over the period of the experiment, and the actual percentage of times it rained was 41.6 percent. That was the only bright spot in the predictions. Of course, this particular prediction average was taken over all four prediction periods. The real question is whether the four-day predictions are as accurate as the one-day predictions.

In too many cases not to be significant, the outcomes of the four-day predictions were worse than the one-day predictions. Indeed, the four-day predictions were out by an average of 28 percentage points, while the one-day predictions were off by an average of 18 percentage points.

As this example suggests, weather prediction has become about as much of a science as it is ever likely to be. The uncertainties produced by chaotic systems can pile up pretty rapidly, not only in days, but sometimes in hours.

Other, major uncertainties surround our fair planet. Has the Earth's magnetic field reversed itself at random moments in the past due to chaos in the circulation of the magma? Were past ice ages, influenced as they were by weather, with disturbances at all scales, also results of chaos? Are the motions of planets and planetoids unpredictable over a time scale that is consistent with their orbital velocities (i.e., hundreds of years)?

Closer to home, are cardiac arrhythmias likely to continue to show up at unpredictable times?

The answer to all these questions seems to be "yes." And there isn't a thing we can do about it. Or is there?

IS THERE A WAY AROUND IT?

It has been suggested that if we can detect chaos in a system, we should be able to do something about it. For example, a visitor to Lorenz's lab remarked that if whole systems could be so heavily influenced by small changes, why not simply make those changes that bring about desired results. Not possible, replied Lorenz. One simply wouldn't be able to predict the overall effect, however profound, of any small change. To see why this is so, suppose we had a sophisticated computer program that applied Lorenz's equations on a global scale and was dead accurate, at least in theory. Assuming that the system were currently on track, we might run the system ahead a day or two to predict the weather, only to find a killer typhoon headed for the Bay of Bengal. Backing up and doing some experiments, we might then discover that a relatively small westerly-directed current of air in southern Italy would avert the tragedy. Accordingly, giant fans in the south of Italy are set blowing to the west.

Strangely enough, the typhoon in the Bay of Bengal turns out to be twice as bad as predicted. What went wrong? It turns out that because the computer only carries, say, 100 digits in its calculations, it missed a whole new scenario in which the typhoon in the Indian Ocean turned out to be a minor tropical storm. It also missed the fact that the Italian experiment would convert the storm into a full-fledged category 5 hurricane.

Is it possible that future mathematical and physical researches will ameliorate the situation? Insofar as our ability to control chaotic systems will depend on our ability to predict their behavior, the answer is "no," with one caveat. Some systems, especially human-made ones such as the wings on modern high-speed aircraft, are not meant to behave chaotically. It might be possible to design systems that avoid the chaotic regime altogether. At least now we know what we're up against.

As for the weather, we might buy one day of extra accuracy after the most strenuous technological effort. After that, forget it.

Math in the Holos

· 5 ·

The Circular Crypt

Unconstructable Figures

> IT IS NOT POSSIBLE, USING ONLY RULE AND COMPASS, TO CONSTRUCT A SQUARE EQUAL IN AREA TO THAT OF A GIVEN CIRCLE.

YOU HAVE BEEN GIVEN a clean sheet of paper, a compass, and an unmarked ruler. On the paper someone has drawn a circle. Can you construct a square, using only the ruler and compass, that has the same area as the circle? If so, you will not only be the first person to solve the problem, but you will also be the first person ever to have contradicted a certain body of mathematical theory, in this case one that implies the feat is impossible.

I must add immediately that the word "theory" appears here in the strict sense. It is not a matter of opinion or conjecture, but an inescapable fact. From the time of the ancient Greeks until late in the nineteenth century, mathematicians and philosophers, not to mention myriad ambitious amateurs, thought that squaring the circle was simply a difficult problem, but not impossible. So many people have tried, in fact, that a distinct pathology has been identified by mathematicians. The chief symptoms of the disease called *morbus cyclometricus* are blurring of vision, sleeplessness, numerous puncture wounds (caused by slippages of the compass), and, of course, circles—under the eyes. There is no known cure.

Since 1882, when the German mathematician Ferdinand Lindemann finally demonstrated the impossibility of the task, "squaring the circle" has become a metaphor for hopeless enterprises.

The story behind the problem has many sides. There is, first of all, the remarkable development of Greek mathematics, beginning in about 600 B.C. and extending to A.D. 350, a span of nearly a thousand years. If Greek mathematics had a soul, it was split between the concepts of number and line, two polarities with no middle ground. The Greeks understood that line and number were, to a certain degree, aspects of the same underlying reality. At the beginning of the Greek millennium, and for nearly a hundred years, they thought that *all* numbers were rational—that is, ratios of integers (or fractions). In this mathematical worldview, the line consisted of points, and every point corresponded to a rational number. In other words, every point on a line lay at a rational distance from a given point. The idea, when you think of it, is really quite beautiful. It was undoubtedly one of the rungs by which Pythagoras and his followers had hoped to ascend to the Olympian ideal. Mathematical knowledge, especially general knowledge, was akin to spiritual development. Unfortunately, what is beautiful isn't always true, Keats notwithstanding.

The first crisis of Greek mathematics was the discovery in about 530 B.C. by Pythagoras, of irrational numbers. His proof that $\sqrt{2}$ could not be expressed by any rational number ranks as one of the great achievements of Greek mathematics, not because it is a difficult result (a schoolchild could understand the proof), but because it stirred the Greek mathematical soul to its depths. The crystalline ideal was shattered. Both number and line were more complicated than anyone had realized. There is no better illustration that mathematics is not "constructed."

Another side of the story concerns the development of what modern mathematicians call the "real line." Essentially the same line that the Greeks pondered, it was found to have undreamed-of structure. Whereas the Greeks had rational numbers and a relative handful of irrational ones, modern mathematicians discovered a deep structure within the irrationals themselves. Beyond the roots of integers such as $\sqrt{2}$ and $\sqrt{5}$, yet containing them, they found the so-called algebraic numbers, which could emerge only as the roots of integer polynomials, both concepts that I will explain presently. If that weren't enough, they also discovered numbers that were not even algebraic, the so-called transcendental numbers, perhaps because they transcended mere algebra, so to speak. One of the newly discovered transcendental numbers turned out to be π, the ratio of the circumference of a circle to its diameter. Our story revolves around π.

Before starting out, there is another, simpler story, one that illustrates the passion this famous classical problem once inspired.

THE π WAR

Thomas Hobbes was a seventeenth-century English philosopher who took an interest in many subjects, from the structure of society to the nature of science. He studied and wrote at a precarious time for England. The monarchy had been temporarily overthrown by Oliver Cromwell, and Parliament reigned supreme. Royalists were greatly alarmed at these developments, and it was during this time, in 1651, that Hobbes, who might be called a radical royalist, published his most famous book, *Leviathan*.

In *Leviathan*, Hobbes asserted that theological entities were immaterial things and therefore outside the realm of rational consideration. He also declared that the British clergy amounted to a "Kingdome of Darknesse." The work embraced a thoroughgoing materialism, on the one hand, while urging the need for the absolute authority of monarchy, on the other. The monarchy would rule the church itself. Even to sympathizers, some of Hobbes's views were hard to stomach. To less sympathetic souls, Hobbes's name was anathema. They called him the "Monster of Malmesbury," after the town where he lived.

In 1628, at age forty, Hobbes became fascinated with Euclid. Hobbes saw in geometry the very essence of scientific reasoning. This was a common view, as geometry had long been considered the best example of what Aristotle called a "science." But Hobbes embraced it almost as a religion, a body of certain knowledge to which nothing could be added except, possibly, an actual solution to the ancient problem of squaring the circle. In 1655 he published a treatise called *De Corpore* (On the Body), which attempted to recast all of science, including mathematics, as the study of bodies at rest or in motion. Although the work was more about meaning than method, Hobbes sought to illustrate the power of his views by squaring the circle. In his solution he used only constructions that, individually, Euclid himself would approve of. The only problem was, they did not actually produce the results that Hobbes claimed.

John Wallis, the Savilian professor of geometry at Oxford, took grave exception to Hobbes's "solution." There was, to be sure, more than mathematics going on here. Wallis was an ordained minister (as were many university appointees in those days) and, while his opinion of Hobbes's demonstration was based on the finding of various logical flaws that doomed the constructions, there can be no doubt that Wallis took a savage pleasure in deflating the pretensions of such a dangerous man. If Hobbes was wrong about squaring the circle, perhaps he was wrong about everything else.

Within a year, Wallis published *Elenchus Geometriae Hobbianae*, a treatise that pointed out numerous technical errors in Hobbes's work, undermining his claims to have squared the circle. As if this weren't enough, he also published a shorter work for more general consumption: *Due Correction for Mr. Hobbes; or Schoole Discipline, for not saying his Lessons right.*

In 1657 Hobbes replied with *Markes of the Absurd Geometry, Rural Language, Scottish Church-Politiks, And Barbarismes of John Wallis Professor of Geometry and Doctor of Divinity.* This work amounted to a general offensive, attacking Wallis on several fronts simultaneously. The bitterness and personal animosity behind these broadsides seemed only to grow with time, guaranteeing a war that would last until Hobbes's death twenty-two years later.

Hobbes did not defend his constructions, perhaps because he realized they contained errors. But he thought the errors were unimportant. Instead, he attacked Wallis where it hurt the most. In 1657 Wallis published his own magnum opus, *Mathesis Universalis*. In it he argued for the primacy of arithmetic over geometry in that all geometrical results could be recast arithmetically. Hobbes attacked the book in 1660 with dialogues in *Examinatio et Emendatio Mathematicae Hodiernae*. As if to rub salt into whatever wounds he may have inflicted on Wallis, Hobbes introduced another solution to the circle-squaring problem in the new book, as well as solutions to two other classical problems bequeathed by the ancient Greeks: duplicating the cube and trisecting an angle.

The war continued until Hobbes's death in 1679. By then, the mathematical world had developed enormously, bypassing the petty struggle in a sense. Descartes' analytic geometry was followed by the infinitesimal calculus developed by Newton and Leibniz. Thinking Euclid the ultimate form of mathematics, Hobbes had reacted to both developments with horror, probably failing to understand the idea of a function or, worse yet, quantities that might be arbitrarily small without being zero. Meanwhile, although mathematicians (those who cared to read the bitter documents) agreed that Wallis had "won" the war, they wondered why he had spent so much time on the struggle.

Squaring the circle had become for both men a kind of mathematical touchstone on which hung much larger issues of social reform and religious philosophy. As well, it might be conjectured that the lure of a shortcut to fame had proved too much for Hobbes. Because Hobbes had already developed something of a following among royalists, Wallis found it necessary to attack Hobbes where he, Wallis, was most competent to do so. In the end, it proved to be Hobbes's jugular.

The π war has been reenacted many times since then. Every year at least one eager amateur submits a magnum opus to a professional mathematician whose heart sinks at the sight of incredible diagrams and scrawled, badly organized thoughts, also symptoms of *morbus cyclometricus*. The mathematician sighs and quietly asks, "Is there any point in mentioning transcendental numbers?"

THE PROBLEM

In the attempt to construct a square that has the same area as a given circle, you are allowed to use only ruler and compass. These tools have a hoary past. The Egyptians, the Babylonians, and the Greeks all used them in laying out the patterns for temples and other public buildings. With the aid of a "ruler," which could be a straight edge made of metal or wood, or even a stretched string, one could produce straight lines. The compass might resemble the modern compass or it might be a string with one end fixed while the other traces a circumference. With it, one can produce circles, arcs of circles, or simply transport distances from one part of a pattern to another.

In his famous *Elements*, the first real mathematics text ever produced, Euclid described ruler and compass constructions for a variety of figures, as well as proving nearly all the geometrical theorems known to Greek mathematics in Euclid's place and time—Alexandrian Egypt, c. 300 B.C.

For example, *Elements* not only showed how to construct a right angle but it also included a proof of the famous theorem of Pythagoras, which states that the square on the hypotenuse of a right-angled triangle equals the sum of the squares on the other two sides.

To construct a right angle at a given point A from a given line L, you may set your compass to any distance, place the point of the compass at A, and describe two arcs that cut L at two new points, B and C, as shown in the following figure.

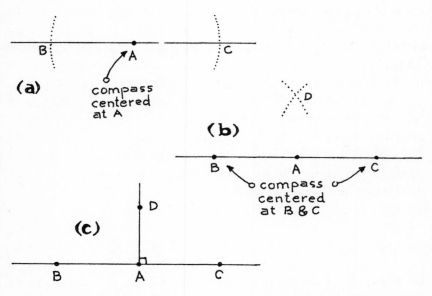

Constructing a right angle by drawing two arcs beside point A (a), drawing two more (b), then joining the intersection point to A (c)

Now place the point of the compass at B and C in turn, drawing two arcs that intersect above A at a fourth point, D. Now simply join A to D, and voilà! There's the right angle at A, with arms AD and AC.

I won't bother to prove Pythagoras' theorem, but I can illustrate it with the right-angled triangle in the next figure. It has a hypotenuse that is 6.5 units long, the remaining sides being 3.9 and 5.2 units long. The bar indicates a unit of length.

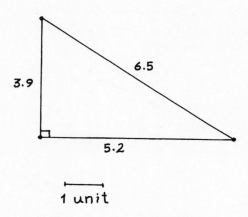

An arbitrary right-angle triangle

Now watch this:

$$(6.5)^2 = (3.9)^2 + (5.2)^2$$

$$42.25 = 15.21 + 27.04.$$

If you square the three numbers as shown, then add the two numbers on the right-hand side of the second equation, they will equal 42.25.

In mathematics the most commonplace observations can have a profound impact. For example, the scale bar is entirely arbitrary. Replace it by any length bar you like, call that one unit, then carefully measure the sides of the right-angled triangle according to the new scale. Then put the new values into the equation. It will still be true. Alternatively, you may keep the same scale bar and draw any other right-angle triangle. The relation will hold for the sides of that triangle as well. One simple equation suddenly becomes true for an infinite number of different right-angled triangles. That's mathematics!

Another wonderful and very useful theorem in Euclid goes back to Thales, the teacher of Pythagoras. In about 600 B.C., Thales discovered his famous theorem of proportional triangles, although we cannot rule

out the possibility that he got the idea from Egyptian priests, whom he
visited on frequent trading trips to Egypt.

Thales' theorem of proportions involves any two triangles that have
exactly the same shape, albeit different sizes. It comes as no great shock
to learn that corresponding sides of the two triangles all have the same
ratios. I have illustrated the theorem in the next figure, where the first triangle has sides of length 1.42, 3.20, and 3.78, while the second has sides
of length 2.13, 4.80, and 5.67.

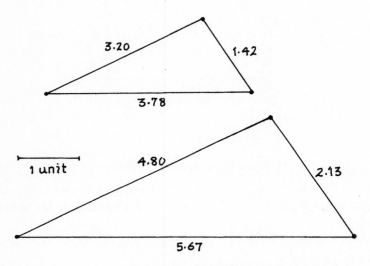

Proportional triangles

To check that the two triangles are similar, merely measure the angles
at each pair of corresponding corners and note that each pair has exactly
the same number of degrees. According to Thales, the ratio of the first
pair of sides, 2.13/1.42, should equal the ratio of the second pair of sides,
4.80/3.20, and, for that matter, also the third pair of sides, 5.67/3.78. A
brief calculation reveals that each of these ratios equals 1.5.

With this modest intellectual equipment I will now attempt to square
the circle myself. I realize that it's supposed to be impossible, but what if
Lindemann and the others are wrong? Impelled by the same egomania
that drives all true circle-squarers, I will begin.

With my compass, I first draw a circle, pristine on a sheet of white
paper. Then, using my ruler, I draw a line through the center and extend
it to meet the circumference of the circle at points A and B.

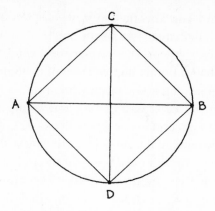

First attempt at squaring the circle

Next, using the construction that I discussed earlier, I will construct a right angle to the diameter at the center of the circle. I then extend the new line to cut the circumference at two new points, C and D.

The four points, A, B, C, and D, form the corners of a square, which I can now draw, using the ruler. Does this square not have the same area as the circle? Oops! Something's wrong. How can a square that is contained wholly within a circle have the same area as the circle? My construction has failed.

Let me try again. I'm sure that if I am clever enough, I can construct the appropriate square. This time I will construct a line at right angles to the diameters at each of the four points A, B, C, and D. These lines will become the sides of a new square, as shown in the next figure.

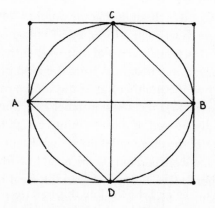

Second attempt at squaring the circle

Alas! This square contains the circle, so it is too large and couldn't possibly have the same area! I will give up (for now), but readers are free to continue this noble project in their spare time. When you at last give up, you will have a keener appreciation of the amazing consistency of mathematics—not to mention writer's cramp.

As we are about to see, squaring the circle is impossible because π, the ratio of a circle's circumference to its diameter, is the wrong *kind* of number. If you try to square the circle yourself, you will find that all your geometrical contortions will only lead you in circles, so to speak. The lengths of every line that you produce will all be numbers of a specific kind. Meanwhile, π lurks outside this realm altogether, grandly indifferent to your efforts.

NUMBER AND LINE

The concept of the line had philosophical and practical implications for the Greeks. Philosophically, the innocent-looking line that Greek geometers drew with ruler and pen was understood to be the temporary and ephemeral earthly embodiment of an abstract and imperishable ideal. That ideal amounted to a continuum of rational lengths. That is, if we mark one point on the line as a reference point, establish what we mean by a unit distance, then the distance from the base point to any other point on the line would always be a rational number. Few people today can grasp the beauty of that ideal. Every point on the line is a rational number, and every rational number is a point on the line. Arithmetic and geometry are unified!

Until Pythagoras, it was thought that the numbers on the line were all rational numbers, such as $3/4$ or -45.52347. Pythagoras went through the professional (and personal) crisis of discovering that some points on the line were not at rational distances from the base point. The specific villain in this case was the number $\sqrt{2}$, or the square root of two, which has the honor to be the first irrational number ever found.

As Pythagoras discovered, if one examines a square with sides one unit long, the length of the diagonal of the square is not rational. In other words, if this diagonal were laid out on the ideal line from its base point, the point at the other end would not "fit" in the ideal line. It would not lie at a rational distance from the base point. As part of the ideal line, the point in question could not belong. Yet there it was!

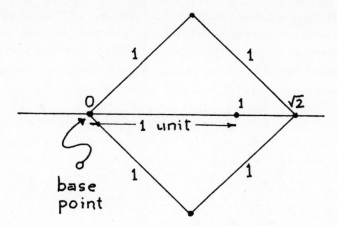

An irrational length on the rational line

By the time the brilliant star of Greek mathematics faded in Ptolemaic Egypt and the Roman Empire rose to preeminence in the Mediterranean, the Greeks had discovered a great many irrational numbers, mainly square roots of integers. They left us with a fundamental dichotomy: real numbers were divisible into two disjoint sets: rational numbers and irrational ones.

Toward the end of the nineteenth-century, European mathematicians had discovered extensions to this scheme. The irrational numbers were themselves divided into two disjoint sets: algebraic numbers and transcendental ones. At the same time, the developing sciences of topology and analysis brought new insights into the structure of the real line. The most dramatic new views of the real line became visible through the twin lenses of density and countability.

The real line is "dense" with numbers. Given any stretch of the real line, no matter how small, we will always find a number there. In fact, even the rational numbers are dense in the real line: between any two rational numbers, no matter how small, we can always find a third rational number. The proof of this fact is very simple. Let a and b be any two rational numbers, no matter how close together. There will always be a third rational number between them, namely $(a + b)/2$. The process of averaging rational numbers can be continued indefinitely. Between the intermediate rational and either of its "neighbors," there will be still another. And another, ad infinitum.

Yet the plethora of rational numbers is not enough to "fill" the real line. Gaps remain. After all, the rational numbers are "countable," meaning that we can enumerate them, 1, 2, 3, and so on, eventually counting every rational number that exists. The standard enumeration of the rationals can be grasped through a simple table, as shown in the following figure.

The rational numbers are countable

It does not take long to realize that with a steadily increasing denominator in one direction and a steadily increasing numerator in the other, the table must contain *all* the rational numbers. By threading the table in the following manner, we can enumerate the rationals, putting them into a one-to-one correspondence with the integers. The enumeration would proceed as follows: $1/1$, $1/2$, $2/1$, $3/1$, $2/2$, $1/3$, $1/4$, and so on.

The irrational numbers, both the algebraic and the transcendental numbers taken together, are also dense in the real line—far denser, in a manner of speaking. The irrationals are far too numerous to even be countable. In fact, if you removed the rational numbers from the real line, you'd hardly notice the difference, so dense are the irrationals.

CONSTRUCTABLE NUMBERS

When I take ruler and compass in hand and make any allowable construction with them, only diagrams of a specific kind will emerge from the process. The crucial question revolves around the points I construct and the distances between them. What are those distances? Is there a property they all have that would help us to understand why squaring the circle is impossible? Here we will engage in mathematics in a more detailed way, beginning to answer the question, then recognizing the emergence of a new concept—the constructable number.

Starting with a long line, a base point, and a point at unit distance from the base point, the distance between the two points is 1. Using the compass, I can produce a point at distance 2 from the base point by merely adjusting the compass to the distance between the base point and the point at unity, then moving it to the unit point and making an arc with the compass to cut the line at a third point, which will obviously be at distance 2 from the base point. I can repeat this performance, establishing points at integer distances from the base point:

    1, 2, 3, 4, 5, . . .

Thus I can "construct" the positive integers as distances within an achievable diagram. What other numbers can I construct? It will be useful to have a name for such numbers. We may as well call them *constructable*.

In the next step of our investigation, we find that any rational number is constructable, thanks to the wonderful theorem of Thales involving similar triangles.

Suppose you give me any rational number, such as 37/22. If you don't mind, I'm going to call this ratio a/b. The construction I provide will use these symbols instead of 37 or 22. As a result, it will be applicable not only to this rational number but to all rational numbers as well.

I'll begin by constructing a line a units long from the base point O to the point A at distance a from O. I will then construct a second line at an angle to the first one and passing through the base point O. The actual angle doesn't matter, as long as it's both acute and constructable. For example, I could construct a right angle, then bisect the angle to achieve an angle of 45°. In any case, the second line will be b units long and end at point B, a distance b from O.

I will now add two lines to the diagram. The first joins point A to point B. The second line depends on a construction in Euclid: to draw a line parallel to a given line and passing through a given point. In this case the second line will pass through U, the unit point on the second line,

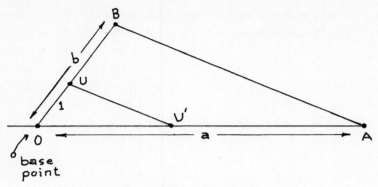

The proportionality diagram

as in the figure above. The second line cuts the line OA at a new point. Call it U′ (pronounced "U prime").

There are two triangles lurking in this diagram. Let me name them: OAB and OUU′. The triangles are similar, technically speaking, because their internal angles are equal in pairs. In fact, the two angles at O are not only equal but also identical. According to Thales, since the triangles are similar, the ratio of OA to OB equals the ratio of OU′ to OU. Writing the relationship symbolically, we have:

$$\frac{OA}{OB} = \frac{OU'}{OU.}$$

Now the ratio OA/OB can be written more simply as *a/b*. At the same time, we know that the distance OU is simply unity or 1 and that any number divided by 1 is simply the number itself. With thanks to simple algebra, we can now write

OU′ = *a/b*.

I have thus succeeded in constructing a line (OU′) that has length *a/b*.

Since it doesn't matter what integers *a* and *b* are, the construction is perfectly general. As a mathematician, I would be entitled (after making the argument more formal, of course) to spell this out as a theorem.

Theorem 1: Any rational distance is constructable.

Sometimes, after a theorem has just been established, a mathematician will recognize that the result is more general than he or she first thought. In such a case there is a new theorem.

Theorem 2: The ratio of any two constructable distances is itself constructable.

The proof of this theorem would use the same argument that established the first theorem. I would simply replace the two integer distances, *a* and *b*, by constructable ones (also called *a* and *b*), also replacing the two integer-length lines in the argument by constructable lines, OA and OB.

So far we have found that all rational numbers are constructable. But as Pythagoras discovered, there are constructable numbers, such as $\sqrt{2}$ that are not rational. The next theorem generalizes this result.

Theorem 3: The square root of any integer is constructable.

This result can be proved by induction, a mathematical technique of inference that uses the structure of the integers to establish a general truth. It says that if a thing is true of the number 1 and if, further, you can establish that whenever it's true for the integer *n*, then it's also true for *n* + 1, then it's true for all integers. Here's the technique in action.

The integer 1 is constructable, since it's the unit length with which all constructions start. The square root of 1 (which is 1) is clearly constructable as well.

Next, we suppose that the theorem is true for the constructable integer *n* and we then try to establish the same thing for the integer *n* + 1. Not surprisingly, we will make an actual construction, a triangle with sides of length 1 and $\sqrt{n}$ set at right angles to each other. The length of the third side, the hypotenuse, can be readily calculated using the famous theorem of Pythagoras.

square of hypotenuse

$$= (1)^2 + (\sqrt{n})^2$$
$$= 1 + n$$

length of hypotenuse

$$= (\sqrt{n+1})$$

It follows that the square root of the number *n* + 1 is also constructable. By the principle of induction, we now know that the square root of any integer is constructable.

Theorem 4. The square root of any constructable number
is itself constructable.

Strangely enough, when I try to construct the cube root of a con-
structible number, I consistently fail. The time has come to cast our net
more widely and take advantage of post-Grecian mathematics. Enter the
algebraic numbers.

The discovery of algebraic numbers marks the first genuine advance
in our knowledge of real numbers since the Greeks. It began, strangely
enough, with a failed attempt to prove Fermat's Last Theorem, a conjecture that many mathematicians had tried to prove since Pierre de Fermat
first stated it in 1639. (The famed "Last Theorem" was not proved until
recently [1998] by the Cambridge mathematician Andrew Wiles.)
Among the better mathematicians to attempt a proof was the German
mathematician E. E. Kummer, who, in 1845, thought he had found a
proof. However, an error was discovered in his reasoning, and Kummer,
anxious to get to the bottom of the mystery, launched an investigation
into algebraic numbers.

Fermat's Last Theorem stated that the equation

$$x^n + y^n = z^n$$

has no integer solutions for any integer n greater than 2. Interestingly,
when $n = 2$, Fermat's expression recalls the Pythagorean theorem

$$x^2 + y^2 = z^2,$$

where x, y, and z might represent the sides of a right-angle triangle. But
if $n = 3$, the equation

$$x^3 + y^3 = z^3$$

has no solutions. More precisely, there are no nonzero integers that you
can substitute for x, y, and z that will satisfy the equation, making both
sides of the equation equal.

The numbers inspired in part by Fermat's Last Theorem were the
"roots of polynomials with integer coefficients." Here is a deliberately
complicated example of such an expression:

$$5x^6 - 17x^5 - 3x^4 + 25x^3 + 53x^2 - 2x - 34.$$

A polynomial in x is any set of powers of x all added together. The
numbers that multiply each power are called coefficients. You can see in

the preceding example that each coefficient is an integer. The roots of this equation are any values of x that cause the expression to take the value 0.

This is tantamount to solving the equation as

$$5x^6 - 17x^5 - 3x^4 + 25x^3 + 53x^2 - 2x - 34 = 0,$$

which I definitely will not try to do. But here is a much simpler example with coefficients 1, 0, and –2:

$$x^2 - 2 = 0.$$

This example is much easier to solve because it asserts simply that

$$x^2 = 2,$$

so that

$$x = \sqrt{2}.$$

It follows that $\sqrt{2}$ is an algebraic number. Meanwhile, the equation

$$x^2 + 1 = 0,$$

also a polynomial with integer coefficients, has a very different kind of root,

$$x = \sqrt{-1},$$

the famous imaginary number that gives us a springboard into an entirely new numerical realm, that of the complex numbers. Using the letter i as shorthand for the square root of –1, we may write every complex number such as 4 + 3i as the sum of a real part, 4, and an imaginary part, 3i. Thus the roots of polynomials with integer coefficients are not always real numbers. They may also be complex numbers, or a mixture of the two.

Recalling that the square roots of constructable numbers are also constructable makes us wonder if all constructable numbers are algebraic. We can show that they are by demonstrating that every constructable number is the root of a polynomial with integer coefficients. In what one of my mathematics professors called "gruesome detail," the proof would take too much space. However, I can lay it out like the plan of a building.

Theorem: Every constructable number is algebraic.

The set of all rational numbers forms a field—that is, a system of numbers that can be added and multiplied, like integers and rational numbers. However, a field must also contain multiplicative inverses; every number, such as 5, must have a partner that, when multiplied against it, leaves unity or 1. Thus $\frac{1}{5}$ is the inverse of 5, since $5 \cdot (\frac{1}{5}) = 1$. The integers, therefore, do not form a field, but the rationals do.

We have already seen that the set of constructable numbers includes all the rational numbers since, for any two positive integers a and b, the ratio a/b is also constructable.

The field R of rational numbers therefore lies entirely within the constructable numbers. The famous nineteenth-century mathematician Évariste Galois developed a method of extending fields by adjoining new elements. We did as much when we noticed earlier that the number $\sqrt{2}$ is also constructable. We make a new field from the rationals, one that contains $\sqrt{2}$, by taking all numbers of the form $a + \sqrt{2}b$. These can be added and multiplied, and the results of these operations can always be put into this form. In fact, for every number like $a + \sqrt{2}b$ there is an inverse, which we may write as $c + \sqrt{2}d$. When we form the product, $(a + \sqrt{2}b)(c + \sqrt{2}d)$, and carry out the multiplication, we get 1. The numbers c and d have the following formulas:

$$c = a/(a^2 - 2b^2) \quad \text{and} \quad d = b/(2b^2 - a).$$

If you substitute these expressions for c and d in the product and then do your algebra correctly, you should get 1 when the dust has settled.

The extension of the field R to include the new number $\sqrt{2}$ may be written in compact notation simply as $R\sqrt{2}$.

The general argument goes like this. Suppose you have a constructable number, r. The construction would have started from a particular point or line, and all subsequent new points (and therefore lines) arrived at in the course of the construction arise in only one of three ways: the new point is the intersection of two lines, of a line with a circle, or of two circles. In the first case, the lines were formed on the basis of points with coordinates in R. The intersection, as it turns out, will also have coordinates in R. But in either of the remaining cases, the coordinates of the new point must be the solutions of a quadratic equation and therefore lying in the extension field $R(r_1)$, where r_1 is the root (like $\sqrt{2}$) of a quadratic equation. As the construction proceeds, new elements are occasionally added. The field builds by a new extension to $R(r_1, r_2)$, then to $R(r_1, r_2, r_3)$, and so on. Since the construction is finite, we end with all the relevant numbers in the geometric construction lying in a field that can be written generally as $R(r_1, r_2, \cdots r_m)$. All the numbers in that field are algebraic because all are roots of polynomial equations. In fact, the polynomials that correspond to constructable numbers have integer coefficients, and all the variables are raised to powers of 2, as in

$x, x^2, x^4, x^8,$ and so on.

Since equations of this form make up a tiny minority in the full set of polynomials with integer coefficients, the constructable numbers are, by the same token, a tiny minority within the algebraic numbers.

Speaking of algebraic numbers, I am reminded of a theorem first proved by German mathematician Georg Cantor. It states that the set of all algebraic numbers is no larger than the set of all integers. It was Cantor, after all, who proved a result quoted earlier—that the rational numbers can be put into one-to-one correspondence with the integers. Using a similar technique, Cantor showed that the algebraic numbers could also be put into such a correspondence. We will meet Cantor again in the next chapter, where he shows that the real numbers *cannot* be put into such a correspondence. This leaves the transcendentals in possession of the field, so to speak. Since transcendentals are by far the most common numbers within the reals, why shouldn't π be transcendental?

There is one final theorem on the road to understanding the impossibility of squaring the circle. It is easy to state but, with a heavy heart, I must confess that the proof is beyond our scope. Although it occupies only a few pages, the explanation of the technical details would take up pretty much the rest of this book.

Theorem (Lindemann): The number π is transcendental.

Even mathematicians find the thirteen pages of Lindemann's proof a bit heavy going. Is there a shorter, simpler proof? Sometimes a proof can be shortened by generalizing. In 1885 the German mathematician Karl Weierstrass made a much more general statement, although his proof was no shorter:

Theorem (Lindemann-Weierstrass): Let $a_1, a_2, a_3 \ldots a_k$, and $b_1, b_2, b_3, \ldots b_k$ be algebraic numbers such that the a_1 are all different and the b_1 are nonzero. Then

$$b_1 e^{a_1} + b_2 e^{a_2} + b_3 e^{a_3} + \cdots + b_k e^{a_k} \neq 0.$$

What does this strange-looking inequality have to do with squaring the circle? Does it show that π is transcendental?

The fact that it does hinges on an extraordinary little formula that links three famous constants—e, i, and π—into a single relationship:

$$e^{i\pi} = -1.$$

Fans of mathematics will recall that e is the base of the natural logarithm, approximately 2.7626, while i is the imaginary number $\sqrt{-1}$, and π, of course, is π—the object of our quest.

Now, if π is algebraic and not transcendental, we may devise a special case of Weierstrass's equation with just two terms. Let b_1 and b_2 both equal 1 and let $a_1 = i\pi$, while $a_2 = 0$. According to the Weierstrass-Lindemann theorem,

$$e^{i\pi} + 1 \neq 0.$$

But this contradicts the previous equation, so π cannot be algebraic.

Since the time of Lindemann and Weierstrass, the theorem has been simplified several times. For example, early in the twentieth century, the German mathematician David Hilbert (whom we will meet again in the next chapter) found a somewhat simpler proof of the theorem. The latest proof of the Lindemann-Weierstrass theorem was published in 1990 by three mathematicians whose nationalities leaven the heavy German presence: Frits Beukers is Dutch, while Jean-Paul Bezevin and Phillippe Robba are French. (Robba, unfortunately, is no longer with us.) The Beukers-Bezevin-Robba proof occupies a mere four pages.

The bottom line for circle-squarers is now obvious. A transcendental number is (by definition) not algebraic, and any constructable number is algebraic. It follows that π, not being algebraic, is also not constructable.

But how, exactly, does the nonconstructable nature of π doom ruler and compass constructions that would square the circle? Suppose you are given a circle of radius 1. The area of this circle is πr^2 or just π (since $r = 1$). If you succeeded in constructing a square of the same area as this circle, you will have produced a square with sides all equal to $\sqrt{\pi}$ in length. This is not exactly the number we now know to be nonconstructable.

However, since you can construct a length of $\sqrt{\pi}$, you can also construct a circle with that radius. The area of the new circle will be πr^2, as before, but this expression now equals $\pi(\sqrt{\pi})^2$, or just π^2. If you are not too exhausted from your construction of the first square, apply your method one more time to the new circle, obtaining a square that has the same area, π^2. The side of this square obviously has length π. Finally, you will have succeeded in constructing π, a direct contradiction of Lindemann's theorem. The original supposition, that you could square a circle of radius 1 in the first place, cannot be correct.

IS THERE A WAY AROUND IT?

I must mention a wonderful construction of Archimedes that does, in fact, square the circle. The only problem is that it's not a ruler-and-compass construction.

In the next figure, I have drawn a circle and a right-angle triangle. The short side of the triangle equals the radius of the circle, and the longer side equals the circumference of the circle. Archimedes proved that the area of the triangle equals that of the circle.

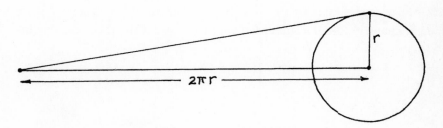

How Archimedes squared the circle

Given the triangle, it is an easy matter to construct a square of the same area. Such a square would have the same area as the circle!

There's only one hitch in Archimedes' scheme. He never explained how to construct a straight line equal to the circumference of a given circle. As Archimedes would freely have admitted, he never solved the problem. Yet his famous construction may have encouraged many people to believe that the task was somehow possible.

R_X for cases of *morbus cyclometricus*: read Lindemann's proof and call me in the morning.

· 6 ·

The Chains of Reason

Unprovable Theorems

⟦
THERE ARE SOME THEOREMS (TRUE MATHE-
MATICAL STATEMENTS) THAT WE WILL NEVER
BE ABLE TO PROVE.
⟧

THE MAIN ACTIVITY of research mathematicians is the search for new theorems. For example, if a mathematician (or anyone else) had ever succeeded in squaring the circle, the following statement would soon have appeared in a leading journal:

"**Theorem:** Pi is a constructable number."

Unfortunately, no one has been able to square the circle. As you may recall from the previous chapter, this limit on our abilities is imposed by a theorem discovered by the German mathematician Ferdinand Lindemann in 1882.

"**Theorem:** Pi is a transcendental number."

Readers will also recall that transcendental numbers are not algebraic and therefore not constructable, either. Pi is forever beyond our reach via standard geometric constructions.

Mathematicians tend to accept such limitations philosophically. After all, a theorem is a theorem, and once it has been proved, that's it. But imagine a limitation on our ability to prove theorems. What if not all theorems are provable?

When mathematicians encounter a statement they think might be true, they call it a *conjecture,* then attempt to prove it. Or they may try to disprove it by finding a counterexample. If the conjecture says that every object with property A also has property B, they may hunt for an example of an object with property A that does *not* have property B. Conjectures are both famous and infamous in mathematics. Our inability to prove them may seem insurmountable until some bright young wizard finds the proof that had eluded everyone else—or finds a counterexample. Conjectures inspire much research and so play a valuable role in the development of mathematics. But should one of the conjectures currently before mathematics happen to be true yet have no proof, then a truckload of young wizards would not suffice to prove it or to find a counterexample. Yet it will be true.

Perhaps an unprovable theorem would involve concepts far beyond our ability to understand. (Who knows?) Or perhaps an unprovable theorem would involve obscure, grotesque, and utterly unappealing content. (Who cares?) Or perhaps an unprovable theorem might be both easy to understand and involve an interesting proposition. Perhaps the conjecture made by the German mathematician Christian Goldbach in 1742 is both true and unprovable. That would be something!

Goldbach's conjecture declares that every even number greater than 2 is the sum of two odd prime numbers. After 250 years, mathematicians have yet to prove the conjecture (which would automatically make it Goldbach's theorem) or to find a counterexample. Yet the conjecture seems to be true. Give me an even number such as 142 and it will not take me long to find two primes that have 142 as their sum. Let's see. How about 59 and 83?

It would be easy to program a computer to search for counterexamples. The program would simply count through all the even numbers and, for each one, check all pairs of primes on a list (maintained by the same program) to see if any two of them summed to the number in question. It would check number after number, mounting ever higher, searching for a counterexample, in effect. Indeed, such programs have been written and, the last time I looked, have confirmed Goldbach's conjecture up to 10^{14}, or 1 followed by 14 zeros. Anyone looking for a counterexample to Goldbach's conjecture would have to look at even larger numbers.

What we used to call "Fermat's Last Theorem" was, until recently, a misnomer, more properly called "Fermat's Last Conjecture." But the misnomer was prescient, in a sense. In 1998, a young Cambridge mathematician named Andrew Wiles proved what we can now, with full justice, call Fermat's Last Theorem: any integer equation of the form

$$x^n + y^n = z^n$$

has no solution for any value of $n > 2$.

If $n = 2$, for example, we can easily find integers x, y, and z, which satisfy the equation. In this particular case, $x = 4$, $y = 3$, and $z = 5$ work quite nicely in that $4^2 + 3^2 = 5^2$. Now try $n = 3$. You won't succeed, according to Fermat (and Wiles).

Will Goldbach's conjecture go the way of Fermat's? Or will we never know? Do true but unprovable theorems exist? If such a thing could be proved, it would be a theorem, to be sure, and a metatheorem, to boot. Such a theorem would have seemed incredible to the Greeks, as well as to the Indian and Arab mathematicians who followed them, no less to the Europeans up to the end of the nineteenth century. Yet this is exactly what the twentieth-century mathematician Kurt Gödel proved in 1930.

Theorem: Some theorems can never be proved.

The actual language of Gödel's theorem is a bit more technical than this. Moreover, there's a back door to the theorem, an escape hatch of sorts: either there exist unprovable theorems or the standard arithmetic is inconsistent. Gödel's theorem is about the "standard arithmetic," a term I will explain later, but which is merely a formalized portion of the mathematics we all learned in elementary school. In short, either there are theorems that our mathematics is simply not capable of dealing with, or our mathematics is itself inconsistent, neither prospect having much appeal for the career mathematician.

From the Greeks to the Europeans—indeed, to all the world's mathematicians in the year 1900—very few things would have been more disturbing than the idea of an unprovable theorem—unless it was an inconsistency within mathematics itself.

The story of this amazing result begins in the year 1900. The setting is Paris, the Second International Congress of Mathematicians. It was an ideal time for one of the world's leading researchers to set the agenda for a new century. The German mathematician David Hilbert challenged his worldwide audience with twenty-three problems. The first two of these would have a profound influence on Kurt Gödel, whose birth lay six years in the future.

THE GHOSTS OF INFINITY

The first problem was to prove the continuum hypothesis; the second was to prove the consistency of arithmetic. We will come back to the second problem in the next section.

On those rare occasions when mathematicians believe a conjecture strongly enough, they dub it a "hypothesis." This does not alter the actual status of a conjecture; it still has to be proved or disproved.

And so with the continuum hypothesis, formulated some sixteen years earlier by another German mathematician, Georg Cantor. As a result of his groundbreaking conceptual invasion of infinity during the years 1871 to 1884, Cantor had formulated a new system of infinite numbers with a strange arithmetic all their own. The first of these was written $\aleph_0$ and called aleph zero. It was the cardinality (infinite, to be sure) of the natural numbers, or counting integers. If the members of any infinite set could be paired off with the numbers 1, 2, 3, . . . and so on forever, that set has $\aleph_0$ members. The second number, $\aleph_1$, aleph one, stood for the cardinality of the set of real numbers. If the members of an infinite set could be paired off with the real numbers, that set would have $\aleph_1$ members.

The continuum hypothesis, as formulated by Cantor, stated that every subset of the real numbers either had cardinality $\aleph_0$ or had cardinality $\aleph_1$. There was nothing in between the two numbers. Every attempt to construct a set of real numbers that was not in one-to-one correspondence with either the integers or with the real numbers met with failure. Thus was born the new field of transfinite arithmetic, with its first two numbers, $\aleph_0$ and $\aleph_1$. However, it was not until 1891 that Cantor was able to prove that the transfinite numbers $\aleph_0$ and $\aleph_1$ were different! To understand Cantor's proof of this result we need one or two mental tools.

The power set of a set is simply the set of all subsets of the set. For example, the power set of the finite set A = {1, 2, 3} consists of eight sets, namely {1, 2, 3} itself, as well as {1, 2}, {1, 3}, {2, 3}, {1}, {2}, }3}, and the empty set. We can write the power set of A as

$$2^A.$$

If A is finite—having, say, n elements—its power set will also be finite and will have 2^n elements, a fact that undoubtedly inspired the notation. If A were an infinite set, its power set would be written in exactly the same way and it, too, would consist of all the subsets of A. Whether finite or not, the power set of a set A always has more elements than A itself, a great many more.

The real numbers, as it turned out, had the same cardinality as the power set of the integers. Writing the integers as Z and the reals as R, we have,

$$R = 2^Z.$$

It is not difficult to explain why this is so, proving a theorem in effect. First, every number can be written as an infinite decimal. If x is an integer, there will be nothing but zeros after the decimal point. If x is a rational number (and not an integer) there will be a finite sequence of digits followed by another sequence that repeats endlessly. I won't bother with integers or rational numbers in this argument, as they already have the same cardinality as Z. (Recall the theorem from the previous chapter, where we showed that the rational numbers could be counted.)

Corresponding to the real number x I will produce a subset A_x of Z by breaking the decimal expansion of x into strings of digits, each of which will become an integer in the set A_x. Since the elements of A_x must all be distinct, I will avoid the possibility of duplicating strings during this process. One way to do this is to select the strings with ever-increasing length—say, 1, 2, 3, and so on. For example, a real number that begins

$$x = 0.702394741023 \ldots$$

might be written as the set

$$A_x = \{7, 02, 394, 7410, \ldots\}.$$

The element 02 warns us that our procedure may not be complete. To prevent any possibility of an integer in A_x beginning with a 0, I will place a 1 in front of every element in the set above:

$$A_x = \{17, 102, 1394, 17410, \ldots\}.$$

It now follows that every real number (of interest) can be identified with a particular subset of 2^Z and that no other real number can possibly become identified with the same subset A_x. In simpler terms, this means that 2^Z is large enough to contain all the real numbers. In the language of transfinite arithmetic, the cardinality of a set of real numbers is no greater than the cardinality of 2^Z.

"Going the other way," as mathematicians say, I must also show that every subset of 2^Z can be identified with a unique natural number. If the set can be presented in some special order—say, with the numbers always increasing—I would simply write them down, with no breaks in between. The final outcome of this conceptual process would be a real number. Compared to an actual detailed proof of the result, my summary is little more than arm-waving, but the essential idea of mapping from one domain (the power set of Z) to another (R) is clear.

The new system of transfinite arithmetic was justified, in part, by Cantor's 1891 theorem, a humble but profound result: the two numbers A_0 and A_1 were different. Specifically, he showed that no pairing or one-to-one correspondence could exist between the real numbers and the integers themselves. His method of proof involved a type of argument that was new to the mathematics of the day but that later would play a key role in Gödel's theorem. His argument used diagonalization, a process that singled out the main diagonal of entries in an infinite table. His proof is simple enough to be presented here.

Suppose that a one-to-one correspondence could be found between Z and its power set, 2^Z. It could then be written as a function f that, for every integer k, would produce a set $f(k)$ of integers. As the variable k ran through the integers 1, 2, 3, and so on, the function f would run through the subsets of Z, all of them, sooner or later. Cantor then examined a very special set that consisted of all the integers z that were *not* members of their corresponding set $f(z)$.

This was a peculiar thing to do. If we made a vast table with the integers down one side and the subsets of Z across the top, every entry of the table would consist of a pairing between an integer and a subset of Z. It would be natural, in such a table, to place the pairs $z, f(z)$ down the main diagonal. That, at least, is the motivation for the name "diagonal argument."

Some of the integers z will appear inside their corresponding subsets $f(z)$ and some won't. If we take the set of all integers z that are *not* members of $f(z)$ and place them in a special set W, we can ask a very serious question about W: To what integer does W correspond under this scheme? If we write that integer as w, we can ask if w belongs to W. If w belonged to W, then w would not be a member of $f(w)$ by the definition of W. But wait! This is a contradiction, since $W = f(w)$. On the other hand, if w were not a member of W, then w must lie in $f(w)$ (i. e., W), another contradiction.

What has gone wrong? Have I just discovered a major inconsistency in mathematics itself? Nothing so drastic. We got into this mess by assuming that there was a function f with the stated property. The assumption must therefore be wrong. There is no one-to-one association between integers and all subsets of the integers. Hence there is no one-to-one correspondence between the integers and the real numbers. It immediately follows that $\aleph_0$ and $\aleph_1$ cannot have the same cardinality and that the distinction between them is real. The question that would immediately suggest itself did so to Cantor. Was there yet another transfinite number between $\aleph_0$ and $\aleph_1$? Cantor thought not. His "continuum hypothesis" remains unresolved to this day.

In his 1900 address to the world mathematical community, Hilbert proposed that resolving the continuum hypothesis was problem one. He also wished to resolve a another question that had been simmering on back burners (and a few front ones) for a decade or more: Could arithmetic be axiomatized in such a way as to guarantee the exclusion of any and all inconsistencies?

CONSISTENCY

By the turn of the nineteenth century, mathematicians had become aware that deep questions attended the simplest-seeming subject: arithmetic. In particular, the attempt to axiomatize arithmetic had led, in some cases, to the recognition of logical anomalies.

For example, in 1888 the Italian mathematician Giuseppe Peano had introduced a set of five axioms that characterized the natural numbers (positive integers). They were nothing if not simple.

1. 1 is a natural number.
2. If a is a natural number, so is $a + 1$.
3. If a and b are natural numbers and $a = b$, then $a + 1 = b + 1$.
4. If a is a natural number, then $a + 1 \neq 1$.

These statements will trouble few readers. From these axioms, with the aid of a fifth axiom that Peano would employ for the deductive process itself, all the properties of the natural numbers could be derived. For example, from these axioms it was possible to prove the associative law that for all natural numbers a, b, and c,

$$a + (b + c) = (a + b) + c.$$

This law justifies something that people do all the time, whether during the mental arithmetic of calculating change at the store or adding up assets: It makes no difference in what order numbers are added.

The fifth axiom would prove troublesome, however. By allowing a set to be arbitrary, it implicitly included a very nasty set that consisted of all sets whatsoever.

5. If A is a set and 1 lies in A, and if for every natural number a in A, $a + 1$ also lies in A, then all natural numbers lie in A.

Known as the principle of induction, this axiom was meant simply as a formal statement of a major tool for working with natural numbers, as it was in chapter 5. If I wish to prove a certain statement about the natural numbers (such as the associative law above), I might proceed by induction. I would apply the fifth axiom by defining the set A to consist of all the natural numbers for which the statement is true. I would then show, as a first step in such a proof, that 1 must lie in A. In the next step, I would assume that an arbitrary number a lies in A and try to prove that $a + 1$ must also lie in A. If I succeeded in the second step, the set A would satisfy the conditions of Peano's fifth axiom, and I could conclude that all natural numbers are in A. In other words, the statement would be true of all natural numbers.

The fifth axiom looked innocent enough. It had merely incorporated an old logical principle that mathematicians had used for centuries in reasoning about numbers. As long as mathematicians using the axiom of induction do not invoke the monstrous set of all sets, they are safe.

This bizarre object seemed to spew out anomalies and downright contradictions. The "set of all sets" sounds almost like a spiritual entity.

In 1893, the German logician Gottlob Frege published the first volume of his *Grundgesetze der Arithmetik,* a rigorous formulation of arithmetic that appealed to the same concept, the set of all sets. The English logician Bertrand Russell was at first unaware of the anomaly and championed the work of Frege, whose unnecessarily difficult book had attracted few readers. But as Frege's second volume was about to appear, Russell discovered to his horror that a severe logical flaw underlay the entire work.

The set of all sets contains some nasty items. For example, there are sets that contain themselves. Why not? Here's a set S. I'll spell it out:

S = {1, 2, 3, S}.

The set S contains the elements 1, 2, 3, and, of course, S. Nothing wrong with that, right? I mean, a set may contain whatever element it pleases, so there's nothing to stop it from being an element, in turn, of another set. Or of itself, for that matter. My vain attempt to render the situation graphically appears here.

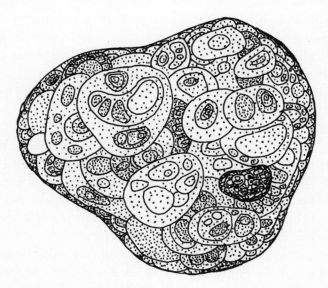

The set of all sets

It is surely no more difficult to conceive of sets that are *not* members of themselves. I will use my favorite nonsense word for such sets, not because I think they are nonsense, but because it illustrates how mathematical terms need have no meaning beyond simply acting as labels. It also happens to underline a certain Germanic flavor in this chapter.

I will therefore call a set that is *not* a member of itself "gzernmplatz." I will go farther and define G to be the set of all gzernmplatz sets. Now I come, as Russell did, to the key question: Is G gzernmplatz? Even an amateur logician would not have to read an account of the discovery. It would have to go something like this: suppose G is gzernmplatz. Then G must lie within the set of all gzernmplatz sets, namely G. Thus G is a member of itself. Whoops! I guess G can't be gzernmplatz after all. It follows that G is not in the set G and, therefore, not a member of itself. So G must be gzernmplatz. Whoops again! Has mathematics suddenly become inconsistent under my feet? Will this seemingly solid bridge of thought disappear under my feet, as in a Road Runner cartoon?

Shut the gates. Avoid such paradoxes by declaring certain infinite sets verboten. The impact on Frege's second volume was devastating, and the *Grundgesetze* was nearly forgotten, even though it contained many good things. For example, Frege had established the consistency of Peano's axioms in a manner that later turned out to be correct in spite of the anomalous set lurking in the works.

Mathematical logic was coming of age as the century turned. There was a growing awareness of difficulties in axiom systems, in the tendency for paradoxes to leap out from the shadows of the subject. It also had been extremely difficult to prove even the simplest of fields, such as arithmetic, consistent. No one knew if serious contradictions might someday appear in our reasoning about numbers.

The dream of a purely logical formulation of mathematics itself, a recasting of the entire field within the emerging area of metamathematics, became the passion not only of Hilbert but also of Bertrand Russell and his English colleague Alfred North Whitehead. In 1910 Russell and Whitehead published the first volume of *Principia Mathematica,* a strict formulation of mathematics, including arithmetic, in terms of a purely logical system that involved both propositions and predicates.

In this context, a "proposition" means a statement that that is either true or false but that contains no variables. For example, the statement "3 is an even number" is a proposition. It happens to be false. A predicate, on the other hand, contains one or more variables. The statement "x is an even number" is an example of a predicate. It might be true or false, depending on the value of x. In addition to predicates, the logical system of the *Principia Mathematica* also employed quantifiers, symbols that indicated the extent of truth of a predicate. For example, $\exists x$ (x is even) says that there exists a number x that happens to be even. As such, it is true. The other kind of quantifier, called universal, asserts the truth of a predicate for all values of its variables. $\forall x$ (x is even) is a universally quantified predicate that says that all numbers are even. This particular predicate happens to be false. In the *Principia,* Russell and Whitehead laid out axioms for logic, for deduction, and for arithmetic. They hoped, ultimately, to include all of mathematics, publishing two more volumes in what turned out to be too wearying a task. They had nevertheless clearly demonstrated that not only could mathematics—or significant portions of it—be reduced to logic, but also that all mathematical truths were ultimately logical truths.

Here was the vehicle that mathematicians could ride in search of consistency. Simply demonstrate that the axioms of the *Principia* could never lead to an inconsistency or contradiction.

In 1923, Hilbert rode forth to the lists with a new program of action that he called *Beweisstheorie,* or the theory of proofs. He published the proposal in a paper and followed up at many talks and conferences to promote the idea. But the current climate of logical uncertainly had rattled him. Speaking to the Westphalian mathematical society in June of that year, he declared the mushrooming paradoxes as "intolerable." He sought a "way of escaping the paradoxes without committing treason" against mathematics herself.

In a nutshell, Hilbert proposed the reduction of successively larger portions of mathematics to a symbolic script—marks on paper, as it were. A proof would amount to a sequence of formulas, each derivable by purely logical (and symbolic) operations from one or more of its predecessors, each leading inexorably to the final formula, which would amount to a statement of the theorem being proved. Following in the footsteps of Russell and Whitehead, he cast his ideas in predicate logic. Hilbert's system began with so-called atomic formulas, the simplest combinations of variables and constants, then proceeded to statements or well-formed formulas that were themselves composed of atomic formulas, logical connectives, and quantifiers of the type discussed above.

An example of a well-formed formula in mathematics would be

$$\forall x \,\exists y \text{ s.t. } (x < y) \,\&\, (x + 1 > y).$$

This rather compact notation may be interpreted as follows:

For all x there exists a y such that x is less than y and $x + 1$ is greater than y.

The symbols $\forall$ and $\exists$ are the same quantifiers we discussed earlier. Their role is to specify, for each variable under their jurisdiction, whether the formula is to be true for all values or at least one of them, respectively. The $\&$ symbol is an example of a logical connective, and the expressions "$x < y$" and "$x + 1 > y$" are examples of atomic formulas.

Such a formula may be interpreted in different ways, depending on what particular mathematical system it referred to. For example, the variables x and y might represent real numbers, or they might represent integers. In the former case the formula is true or "satisfiable" because no matter which real number x you choose, there is always another real number y that lies between x and $x + 1$. However, in the latter case, it is not true. If x and y are integers, y can never lie (strictly) between x and $x + 1$. It obviously makes a difference what area within mathematics such formulas are supposed to describe. The whole point of predicate logic was that it amounted to a language in which axiom systems and theorems for various areas of mathematics could be expressed. This language became the focus of interest in the newly emerging field that Hilbert called "metamathematics." It would interest young Gödel profoundly.

THE TROUBLEMAKER

Kurt Gödel was born in 1906, the second son of Rudolph and Marianne Gödel in Brno, a city in the Moravian part of Czechoslovakia. Although living in Czechoslovakia, the Gödels always considered themselves German. Rudolph Gödel was a successful businessman in the textile industry. The Gödels lived quite comfortably, and the children were denied little. In a privileged but highly structured household, young Kurt flourished academically, in spite of a somewhat troublesome constitution. He received high grades throughout his primary school years and his years at the *gymnasium,* or high school. In spite of his strong interest in mathematics, however, Gödel usually scored higher grades in his other subjects. He graduated in 1924 and was sent to study at the University of Vienna.

Gödel began in physics but found himself increasingly attracted to mathematics. While taking courses in physics, he would read the classic works of Euclid, Euler, and others. Gödel examined mathematics from a philosophical point of view, reading Kant and Russell as well. Indeed, his foremost interest in mathematics must have been sparked even more strongly by attending a weekly seminar conducted by the philosopher Moritz Schlick. In the academic year 1925–1926 Schlick focused on Bertrand Russell's *Introduction to Mathematical Philosophy*, later switching to Wittgenstein's *Tractatus*. Sometime that year, Gödel met one of the jewels in Vienna's mathematical crown, Hans Hahn.

Famous for his contributions to a variety of mathematical fields such as the calculus of variations, set theory, and analysis, Hahn had recently turned his attention to the foundations of mathematics. It was Hahn who had brought Schlick to the University of Vienna. It may even have been Hahn who suggested to the young Gödel that he attend another special weekly seminar, an invitation-only affair. It was at this seminar, later to be called Der Wiener Kreis (the Vienna Circle), that Gödel heard the crucial issues and logical questions of the day discussed. Among the brighter lights of the seminar who would influence Gödel was Rudolph Carnap, who viewed the foundations of mathematics as being largely a question of syntax. Although Gödel disagreed with this view, he found Carnap immensely stimulating and probably took his course on the foundations of arithmetic.

By 1927, Gödel was hopelessly involved in mathematical issues and questions. The Vienna Circle had brought him to the heart of difficult and important mathematical questions. He read, and he attended lectures. He walked the streets of Vienna alone or with colleagues. He sat for hours in various coffeehouses discussing mathematics. But this was Vienna, and the young Gödel could hardly refrain from discussing areas of wider interest, even with his older colleagues: theater, art, social reform, even spiritualism. In 1928 Gödel had finished his undergraduate work and by 1929 was already hard at work on his Ph.D. thesis, an attempt to show that predicate logic was complete.

In 1929 Hilbert and his colleague Willheim Ackermann had published their *Grundzüge der Theoretischen Logik*. In his thesis, Gödel succeeded in proving that the logical system suggested by Hilbert and Ackermann was complete. Gödel showed that every valid formula (true expression) was derivable within the system. He began by reducing the problem of proving completeness to showing that each formula within Hilbert's system was either satisfiable or refutable. He established the latter result, in turn, by using induction.

It was an impressive performance from someone so young. But the result surprised no one. Everyone expected Hilbert's system to be complete.

THE TROUBLE

The "habilitation" denotes a second hurdle that had to be leaped by all aspiring academics in Continental universities. It was not enough to write and defend a thesis. If one expected employment at an institution of higher learning, one had to publish something of note *after* the thesis. For his habilitation paper Gödel chose to work on Hilbert's second problem, that of showing the consistency of arithmetic. It would undoubtedly be more difficult than the proof that predicate logic was consistent and infinitely more difficult than the proof that propositional logic was consistent. That result had been achieved by Emil Post, a mathematician at New York's City College in 1921.

The propositional calculus is the simplest form of logic, essentially a subject first codified by Aristotle in the fourth century B.C. The "propositions" are merely symbols such as a, b, and c, which stand for fixed statements that contain no variables. Two propositions could be conjoined logically by either the "and" operator or the "or" operator. For example, $a \lor b$ represents two propositions connected by an or symbol, $\lor$. The two propositions a and b might mean anything:

    a = "John drives his car."
    b = "John walks."

The new proposition $a \lor b$ means "John drives his car or John walks," not terribly exciting and not always true. John, for example, might be sitting at home reading a book. A propositions might be negated, as in ~a, which means "not a" or "it is not true that a." Thus if a had the interpretation above, ~a would mean "John does not drive his car."

In their classic *Principia Mathematica*, Russell and Whitehead gave axioms for the propositional calculus. Instead of the "and" connective, however, they used implication, symbolized by the arrow, →. The expression a → b, which reads "a implies b," is understood at the outset to be logically equivalent to ~a ∨ b.

1. $(p \lor p) \rightarrow p$
2. $p \rightarrow (p \lor p)$
3. $(p \lor q) \rightarrow (q \lor p)$
4. $(p \rightarrow q) \rightarrow [(p \lor r) \rightarrow (q \lor r)]$

Each axiom in this system seems either harebrained or mildly insane. Axiom 2, for example, says that if a proposition p is true, then either p is true or p is true. There is, in any event, very little to argue with in the axioms.

Russell and Whitehead had shown how, with the addition of two rules for manipulating propositions, one could arrive at any theorem in propositional logic. One wrote down a sequence of expressions, each one an axiom or a new expression derived from earlier ones in the sequence by application of either of the two rules.

The *rule of substitution,* as applied to a logical expression, allowed one to replace all occurrences of a propositional symbol p by any expression that occurred earlier in the sequence. This was a reasonable rule because if p stood for any proposition at all, it must also stand for the propositions one could build up within the system.

The rule of detachment enabled one to detach implications. Given that the expressions P and P → Q have both occurred earlier in the sequence, one could now add Q as a new expression. This rule was also reasonable because if a thing P is true and if an implication with P as its premise is also true, as in P → Q, then surely the implicand Q is true as well.

To demonstrate this system in action, I will prove a well-known theorem in the propositional calculus.

Theorem: For all propositions p and q, the statement $p \rightarrow (\sim p \rightarrow q)$ is always true.

This is not a very exciting theorem at first glance, but it looks a little strange. It says that p, regardless of its truth value, always implies that ~p implies q. Does it make sense? Since it is true of all possible propositions, we may suppose that

p = "John drives his car."

and

q = "Roses are blue."

The theorem I have just quoted may then be applied to these propositions as follows: If John drives his car, then it is true that if John does not drive his car, roses are blue. The point is that if "John drives his car" is true then "John does not drive his car" is false, and from an untrue statement anything can be deduced, even that roses are blue.

The theorem may be proved in three steps by appealing to the equivalence of $p \rightarrow q$ with $\sim p \vee q$.

1. $p \rightarrow p \vee q$ (axiom 2)
2. $p \vee q$ is equivalent to $\sim p \rightarrow q$ (equivalence of expressions)
3. $p \rightarrow (\sim p \rightarrow q)$

The consistency of propositional logic may now be proved by supposing that it is possible to derive two contradictory statements, T and $\sim$T, from the axioms. Suppose then that we have discovered a disastrous proposition T such that T and $\sim$T are both true. We may use the principle of substitution, replacing the p of our theorem with T:

$T \rightarrow (\sim T \rightarrow q).$

Since T is true, we may use the rule of detachment to establish that

$\sim T \rightarrow q$

is also true. But $\sim$T is also true, and we may use the rule of detachment once again to show that

q

is also true. But q can be any proposition whatever. The assumption of a contradictory pair of propositions had therefore led to the conclusion that all propositions are true, including the negations of the axioms themselves. It follows from this contradiction that no such pair of contradictory statements or propositions can be derived within propositional logic, and the foundations are secured. It is consistent.

This example also serves to illustrate an important distinction between two kinds of reasoning. You will notice that our treatment of propositional logic involved proofs within the system and proofs outside the system. The derivation of the theorem $p \rightarrow (\sim p \vee q)$, like all theorems of propositional calculus, was proved within the system by applying the rules of substitution and detachment, but the proof of consistency was proved outside the system. Our reasoning was no less formal, but there was no way to express it as a sequence of propositions, each derived logically from previous propositions in the sequence. In short, there was no way to express the assumption about T. The distinction is fundamental to metamathematics. We will see it made again when we consider Gödel's amazing theorem. We will also see how Gödel sneaked around it.

It is still a long way from showing the consistency of propositional logic to that of predicate logic when applied to arithmetic. For one thing, propositional logic is silent on the subject of arithmetic. Numbers cannot be expressed within it.

Gödel began by trying to prove that arithmetic, as expressed in the logical framework of the *Principia,* was consistent. In seeking to express the consistency of arithmetic, however, he discovered that he could express this consistency within arithmetic itself. Not only that, but this very expression led directly to a theorem that could not be derived within the logical system of Russell and Whitehead.

After setting up the axioms of a predicate logic that embodied what is called the standard arithmetic, Gödel drew up a list of all the symbols used and assigned a special code number to each symbol, as shown in the following table:

SYMBOL	CODE NUMBER	SYMBOL	CODE NUMBER
0	1	x	9
s	2	l	10
+	3	⌐	11
X	4	&	12
=	5	∃	13
(	6	∀	14
)	7	→	15
.	8		

The symbol that resembles the letter L rotated on its head stands for *implication* in the predicate calculus. Thus A ⌐ B means simply that A implies B or that B can be deduced from A. The symbol s is the successor function. When applied to a natural number, it yields the next number in sequence. It might be asked why just one variable, *x,* would appear to be allowed for in a system that would surely have to accommodate many variables. However, the lone *x* could readily be made to serve in this capacity by using subscripts, in effect. The subscripts used in unary notation consist of ones. Thus x_1 would be one variable, x_{11} another, x_{111} yet a third.

Any formula developed in the course of a mathematical investigation within this system could have each of its symbols replaced by a number. Because each formula would have to be expressed by a single, unique number, Gödel would need a way of boiling all the numbers that represented a given formula down into one number. Somehow that number would have to encode not only all the number symbols of the formula simultaneously, but also would have to encode their order.

Gödel achieved this trick by using consecutive prime numbers such as

$$2, 3, 5, 7, 11, 13, 17, 19, 23, 29, 31, \ldots$$

each raised to a certain power. Which power he used would depend on the position of the symbol in the formula. If a symbol such as *x* appeared as the seventh symbol in a formula, for example, it would be represented by a 9 (the symbol for *x*) raised to the 17th power (the 7th prime in the sequence).

Thus if Gödel wanted to translate the formula (axiom)

$$x_1 + sx_{11} = s(x_1 + x_{11})$$

into a single number, he would first replace every symbol by its numerical code:

$$9, 10, 3, 2, 9, 10, 10, 5, 2, 6, 9, 10, 3, 9, 10, 10, 7.$$

In this case he would obtain the seventeen integers listed above. Next he would raise the first seventeen prime numbers to these powers and multiply them all together, the raised decimal points representing ordinary multiplication.

$$2^9 \cdot 3^{10} \cdot 5^3 \cdot 7^2 \cdot 11^9 \cdot 13^{10} \cdot 17^{10} \cdot 19^5 \cdot 23^2 \cdot 29^6 \cdot 31^9 \cdot 37^{10} \cdot 41^3 \cdot 43^9 \cdot 47^{10} \cdot 53^{10} \cdot 59^7.$$

The resulting number is huge but finite. The encoding procedure could be specified as a finite process that would, given enough time, serve to express any formula whatever by an integer, no matter how large. It was recognized by the early metamathematicians that in explorations of the infinite, the best way to stay out of trouble was to use only finite tools. The finitary encoding employed by Gödel was just one example of what mathematicians of the day called "recursive," a property that would become increasingly important as computers came into general use a few decades later.

Gödel, by the way, did not actually go to the trouble of converting all his formulas and rules of logic into these vast numbers. He merely showed *that* it could be done by showing *how* it could be done.

Consider, for example, the proof of a certain expression in the logical system employed by Gödel. It would consist of a finite string of formulas. The initial formula would represent an axiom, and all subsequent formulas would represent either axioms or formulas that were obtained from earlier formulas in the sequence by the deductive rules of the system, not unlike the manner in which we proceeded with propositional logic. The last formula in the sequence would be the theorem that the sequence proved. Such a proof could, of course, be encoded by Gödel into a single integer (nowadays called its Gödel number).

Conversely, given any positive integer whatever, there was a finite procedure for producing either a corresponding formula or sequence of formulas, complete nonsense, or nothing at all, depending on what symbols were encoded (or not) and in what order. Thanks to an important theorem within arithmetic itself, any positive integer can be written as the unique product of primes, which, when grouped and placed in consecutive order, would represent the given number uniquely. The powers of the primes could then be translated directly into the corresponding symbols—unless they happened to be larger than 15. The fact that an arbitrary integer might not represent a formula was not a problem. But when it did, Gödel knew that no other integer could represent that formula.

Gödel had thus developed a language of sorts, one that linked two seemingly different universes. On the one hand, there was the universe of formulas; on the other, the Gödel numbers, a system of huge integers that each represented one of these expressions. The very process of thought, as embodied in the formulas, had been reduced to a cloud of numbers.

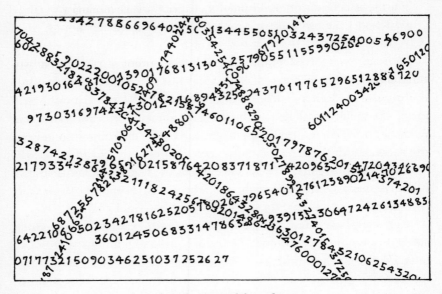

Thoughts as Gödel numbers

Among the formulas that Gödel showed how to construct were ones that would be typical of theorems and proofs. He had shown how formulas could express the process of checking that a proof really was a proof.

The Gödel numbers were all, of course, natural numbers, and as such formed part of the standard arithmetic. This meant that he could construct statements about the Gödel numbers, just as he could for ordinary integers. Within the standard arithmetic, in other words, he could frame predicates that took the Gödel numbers as their subject matter, so to speak. Of special interest is a rather complex predicate that I will write in a simple symbolic form:

Proof(x, y, z).

I have used the traditional variable names instead of x_1, x_{11}, and x_{111} to make the path into Gödel's mind a little easier to follow. The interpretation of the predicate called "proof" depends on knowing that X is a sequence of formulas that allegedly prove something, while x is the Gödel number of the proof X. Another formula, Y, has only one variable, and y is its Gödel number. With these elements in mind, the predicate can be described as saying,

x is the Gödel number for the proof X of a one-variable formula Y
with Gödel number y and with the integer z substituted into it.

In other words, Gödel number aside, X is a proof of Y, with z substituted into it. If the formula Y is true for this value of z, and if the system is complete in that every theorem can be proved in it, the proof X will exist, x being its Gödel number. In this case, of course, the formula Y will be the last formula in the string represented by X. Under these conditions the predicate called "Proof" will be true. However, one cannot simply write any numbers one chooses for x, y, and z and expect the predicate Proof to be true. It would in fact be true only for the sparsest imaginable distribution of such integers.

The expression "Proof(x, y, z)" does not belong to the system under study, but to the metalanguage in which truths about the system are expressed. Yet this shorthand, "Proof(x, y, z)," refers to a series of formulas that express what it means for X to be a proof of Y with z substituted into it, an encoding, if you like, of the mental machinery required to check such a proof. For example, the actual expression would have predicates that checked that each formula in the proof sequence was derivable from ones earlier in the same sequence. To spell out the proof-checking procedure, as long as it amounted to a finite process, was tantamount to an actual check. All that machinery within the logical system, along with the proof X of the formula Y, amounted to an extremely long, but finite, string of formulas. Consequently that string would itself be a formula and would have its own Gödel number.

The next and most important step that Gödel took was to realize that the formula Y referred to in the predicate Proof(x, y, z) could have its own Gödel number, y, substituted into it, instead of the more general variable z. In other words, Gödel's attention now focused on the predicate

Proof(x, y, y).

In this form the predicate is true if x is the Gödel number of a proof that the formula Y is true when its own Gödel number y is substituted into it. In other words, the predicate automatically symbolizes all formulas Y within the system that happen to have a proof X when y (the Gödel number of Y) is substituted into Y. This seems an odd thing to consider.

At this juncture, Gödel's breath may have caught as he sensed himself on a collision course with common sense. What he did next was to form a new predicate that denied the existence of such a proof:

$\sim \exists x \, \text{Proof}(x, y, y)$.

This expression was all that Gödel looked at while working outside the system. It formed an element of the metalanguage that he used to reason about the logical system under examination, the one that contained the standard arithmetic of integers. Yet by merely appending the same logical symbols to the extremely lengthy expression inside the logical system, the one represented by "$\text{Proof}(x, y, y)$," he now had a first-class anomaly on his hands.

The new expression denied the existence of a proof X (with Gödel number x) that the formula Y would be true with its own Gödel number substituted into it. Yet this new predicate, considered as a lengthy expression within the system, would have yet another Gödel number all its own—say, g.

What, asked Gödel, is the status of the predicate $\sim \exists x \, \text{Proof}(x, g, g)$? This expression asserts that there exists no proof of the predicate symbolized by g, namely the predicate $\sim \exists x \, \text{Proof}(x, y, y)$. If $\sim \exists x \, \text{Proof}(x, g, g)$ were true, then no proof could exist. If $\sim \exists x \, \text{Proof}(x, g, g)$ were false then the expression

$\exists x \, \text{Proof}(x, g, g)$

would be true and a proof would exist. But a proof of what? It would be a proof of $\sim \exists x \, \text{Proof}(x, y, y)$, because that's what g stood for. But a proof of this predicate would hold for all possible values of y, including g, leading Gödel to the inevitable conclusion that $\sim \exists x \, \text{Proof}(x, g, g)$ was indeed true.

Here was the crunch. If the predicate $\exists x \, \text{Proof}(x, g, g)$ were both true *and* false, then his logical system—and therefore the standard arithmetic—was inconsistent. The only way out was to assume that the statement symbolized by g was true. And that statement had no proof.

At the time of his discovery of the famous incompleteness theorem, Gödel's own number (age) was twenty-six. Perhaps because he was so young, because he was not then an established mathematician, news of the result did not exactly spread like wildfire. At the same time, the colleagues and contacts to whom Gödel communicated his result most frequently expressed confusion rather than admiration. And yes, resistance grew.

IS THERE A WAY AROUND IT?

Unless someone were to find a subtle error in Gödel's proof of the incompleteness theorem, there are only two alternatives: Either mathematics is inconsistent, harboring contradictory theorems, or mathematics is incomplete, harboring unprovable theorems. To illustrate this stark situation, we examine the aftermath, so to speak, the slow process of acceptance within the mathematical world.

In August 1930, Gödel met his colleague Rudolph Carnap to plan a trip to a conference in Königsberg (today Kaliningrad) in one month's time. Quietly announcing his result, Gödel was surprised to discover that Carnap did not quite follow the argument. At the conference in Königsberg that September, Gödel's presentation was relegated to a twenty-minute "contributed" session near the end of the conference. This was not because conference organizers wished to suppress the result, but rather that they simply hadn't heard about it. Why should such a young mathematician, barely past his thesis, receive any more prominent a platform?

Gödel himself was reticent and retiring when it came to self-promotion, He believed that the mathematical world would eventually catch up to him. He was right, but the process was slow. It began almost immediately, however, with the attendance at Gödel's Königsberg presentation by no less than John von Neumann, a world-renowned mathematician who had published in a great variety of areas and who took a particular interest in the new metamathematics, having spent long hours wrestling with the consistency problem himself. After the talk, von Neumann presented himself to Gödel, congratulating him and inquiring further into the stunning new result. Once he fully understood what Gödel had done, von Neumann suffered a slight fit of pique. He himself had worked to near exhaustion, but in the wrong direction. Yet his admiration was genuine, and his desire to promote Gödel's new result was sincere.

Hilbert, who also attended the conference, apparently was completely ignorant of what was going on. His own contribution, a major speech titled "Naturkennen und Logik" (Logic and the Understanding of Nature), labored along the same road he had first mapped out in 1900.

For the mathematician there is no ignoramibus, and, in my opinion, not at all for natural science, either. . . . The true reason why [no one] has succeeded in finding an unsolvable problem is, in my opinion, that there is no unsolvable problem. In contrast to the foolish ignoramibus, our credo avers: We must know. We shall know.

In a further touch of irony, the conference organizers omitted Gödel's presentation from the conference summary.

Hilbert did not become fully aware that Gödel had dashed forever his hope of an impregnable mathematical fortress until the following January. Paul Bernays, a close mathematical associate of Hilbert, informed him of the result after obtaining a preprint from Gödel. Hilbert apparently was quite upset about the new theorem and probably somewhat depressed as well. After all, his credo that "we must know" had just been shattered by the knowledge that unless mathematics was inconsistent, there would be some things that we would never know, namely, which conjectures might turn out to be unprovable. Ultimately he swallowed his disappointment and reluctantly admitted that things had indeed changed. Bernays, meanwhile, found the new result confusing.

In March 1931 a strange paper by Gödel appeared in the *Monatshefte für Mathematik und Physik* titled "Über Formal Unentscheldbare Sätze *Principia Mathematica* und Verwandter System I" (On Certain Difficulties of Proof in the *Principia Mathematica* and Related Systems). The new paper slowly made its way into the collective consciousness of the mathematical world. The following September Gödel attended a meeting of the German Mathematical Union in Bad Elster. By now news of the result had made the rounds, and Gödel met his first real opposition in the person of Ernst Zarmelo, the mathematician who had first axiomatized set theory, a subject intimately related to logic.

When colleagues proposed getting Zarmelo to lunch with Gödel, Zarmelo at first refused on a suspiciously wide variety of grounds: he didn't like Gödel's looks; he couldn't walk that far; if he attended the lunch, there wouldn't be enough food to go around. His subsequent hour with Gödel seemed outwardly pleasant, but Zarmelo held such different views of logic that he did not fully grasp the import of the incompleteness theorem. Soon after the conference ended, Zarmelo wrote Gödel that he had discovered an "essential gap" in the theorem. Gödel replied that what appeared to be a gap had in fact been filled later in the paper. Zarmelo, unfortunately, continued to misunderstand the proof of the result, but Gödel did nothing more to disabuse him of this condition. Gödel shrank from controversy of any kind.

Other notables also found the result difficult to understand, including the mathematical philosopher Ludwig Wittgenstein and even Bertrand Russell, who expressed gratitude that he no longer worked in mathematical logic. By the mid-1930s there was barely a mathematician alive who did not know of Gödel's theorem and its philosophical implications for mathematics. Most preferred to ignore the first horn of the dilemma raised by Gödel's theorem, that mathematics was privately plagued by monsters of inconsistency.

Even assuming that mathematics was consistent, it would never be quite the same. The mere possibility of unprovable theorems has added a third potential outcome (or nonoutcome) in the pursuit of conjectures. Someone will find a proof, someone will find a counterexample, or no one will find anything.

· 7 ·

The Computer Treadmill

Impossible Programs

THERE ARE SOME YES/NO QUESTIONS, PER-
FECTLY WELL DEFINED, THAT NO COMPUTER,
NO MATTER HOW FAST OR POWERFUL, CAN
ANSWER.

IN TODAY'S ANYTHING-IS-POSSIBLE media climate, many people have
come to believe that there is nothing that computers can't do. But, as we
shall see in this chapter and the next, there are some very important
things that computers cannot do. Moreover, we cannot even conceive of
an overall design that would make computers any more powerful,
whether they are massively parallel machines, quantum computers, DNA
computers, or what have you. Each of these possibilities may make for
faster computers, but uncomputable means uncomputable. Speed means
nothing in this realm. You might even have a computer that doubles its
processing speed every second, but it will make no difference.

There is a fundamental reason for this limitation. All computers are created equal in a very important sense: once a computer (of whatever type) becomes fully programmable, it arrives at a plateau beyond which there can be no progress. By "fully programmable" I mean simply that the computer in question can follow a program in a language that includes instructions for storing and retrieving numbers, as well as performing basic arithmetic on them. Most programming languages go well beyond these simple operations, of course, but it makes no difference to the powers that we will explore in this chapter.

Any fully programmable computer can simulate any other such machine, and this simply means that what one can do, all can do, albeit with greatly varying speeds. Conversely, should there be a task at which one computer is doomed to fail, no matter how long it is given, so will all computers fail. This limitation has a curious hole in it, however. It is not really a theorem, but a conjecture. Even if this thesis should turn out to be incorrect, the limits on computers as we understand them remain. It would just mean that there's a completely different kind of computer that we have not yet imagined, one that is able to surmount these limitations.

The conjecture is known as Church's thesis, named after the American logician Alonzo Church. It is almost certainly true for a strangely profound reason. All attempts to arrive at a definition of what it means to compute something seem to result in equivalent machines. There are three famous results, each discovered in the mid-1930s, each describing a computational scheme that is wildly different from the others. All three schemes turn out to be the same. They describe exactly the same class of functions. Their only shared property is that they proceed in steps of one kind or another.

For the past hundred years, mathematicians and logicians had become increasingly aware that the structure of mathematics itself was a fit topic of research. It became increasingly important to have a description of mathematical reasoning, not necessarily the creative side, when insights come in a blinding flash, but in the plodding, step-by-step fashion in which the mathematics must be written out (and checked) to be sure it is correct. This process, captured by the formalisms of Hilbert, Gödel, and many other mathematicians, had a parallel in algebra itself, wherein one formula leads to another through algebraic manipulations.

Words for the process of step-by-step development of an idea abounded: "mental process," "effective procedure," "algorithm." The latter came to be most commonly used. There is no definition for the word "algorithm" because it is normally used in an informal sense. But it corresponds to the notion of any process that (a) is definite and clear, (b) proceeds by steps, and (c) eventually terminates with an answer. Church's thesis says, in effect, that as soon as one tries to define the word "algorithm," the definition leads to a computational scheme that is logically equivalent to all the other schemes so far proposed. Perhaps there is no way to prove Church's thesis, only to disprove it. This is a peculiar situation because we do not know whether someone will someday come up with a description of an algorithmic process that goes beyond the uniform plateau we call "computing." Until then we shall continue to believe it.

The rest of this chapter will elaborate these simple remarks in a way that takes us into the heart of what it means to compute. Our focus will be the *Entscheidungsproblem,* a formidable German word that means "decision problem." There is not one decision problem, but many. Each decision problem calls for nothing more than a "yes" or a "no" answer. A decision problem for integers would be to decide, for an arbitrary integer, whether it is prime or not. An algorithm exists for this decision problem, so the problem is decidable. A decision problem for the logical formulas studied by Gödel (see chapter 2) would be to decide, for an arbitrary formula, whether it had a proof or not. As Gödel showed in effect, there is no algorithm to solve this decision problem, so the problem is undecidable.

In what follows, we shall witness the conceptual birth of the modern computer in a world of abstract conceptions that seems, at first sight, far removed from the world in which these amazing machines now operate.

THE TURING MACHINE

Alan Turing was a British mathematician who is perhaps best known for his work during World War II for British Intelligence. He devised a specialized machine, an early fixed-purpose computer, that analyzed intercepted ciphers from German U-boats.

But his greatest work was unquestionably in both the theoretical and
practical aspects of computing. The vehicle of his genius was the Turing
machine, a conceptual device so disarmingly simple that even a child
could understand it.

Here then is a Turing machine, shown in diagrammatic form:

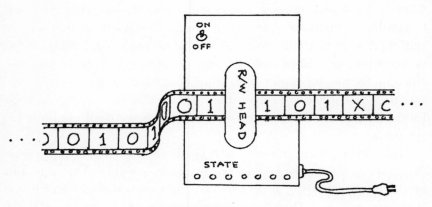

Turing machine with tape

The Turing machine consists first of an infinitely long "tape," which
consists of cells, each cell capable of holding one symbol at a time. For
example, the tape in the figure shows three symbols written in the cells:
0, 1, and X. At any moment, the machine scans a particular cell. It may
replace the symbol it finds in the cell by another, or it may leave the cell
as it is. At the next moment, it may move to the next cell to the right,
the next cell to the left, or not move at all. And that's it—almost.

To guide its actions, each Turing machine comes equipped with a
state-transition table. Each Turing machine has a finite number of states,
and at any moment it must be in one of them. The world of a Turing
machine thus consists of two things: its present state and the particular
symbol it happens to be reading on its tape. When it looks up this combination in its state table, it sees an instruction about what symbol (if
any) to replace the symbol it is scanning. The instruction also tells the
machine which cell to move to next (left, right, or same) and what state
to enter for the next computational cycle.

Here is an example of a state table for a particular Turing machine,
the one that appears in the figure above.

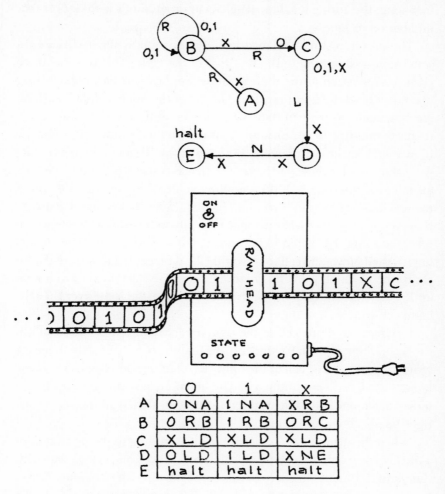

Turing machine state table and transition diagram

This particular machine multiplies a binary number by 2. It does this by starting at the left-hand X marker, then proceeding to the right-hand X marker. It replaces the right-hand X by a zero, then moves one cell farther to the right, where it now writes the X in its new position.

It's not really a very impressive feat. It has simply doubled the binary number by adding a 0 to the end of the string that represents the number on the tape. But it is fully general within the framework of that simple task. Given any binary number (suitably framed by markers), no matter how long, the Turing machine will sooner or later quit with double that number on its tape.

The state-transition table works like this: initially the machine is in state A, corresponding to the first row of the table. This row has three parts, one for each of the three symbols used by this particular Turing machine. Thus we interpret the first row of the table as follows: if the machine happens to be in state A and is examining a cell containing a 0, it enacts the triple 0NA, meaning that it writes a 0 in the cell (ignoring it, in effect), makes a nonmove (represented by N), and remains in state A. In short, under these particular circumstances, as long as the machine examines a 0 in state A, it does nothing at all. The same thing is true if the machine happens to be examining a 1. Under the column labeled X, however, we discover what the machine will do if it sees an X in the cell it is examining. The triple XRB means that it will write the X (i.e., ignoring it), then move one cell to the right (R), then enter a new state (B). In short, the machine won't do anything unless we start it over a cell with an X in it. According to my description, we must place it over the left-hand cell with an X in it.

Having moved one cell to the right, our machine encounters a 0. In the second row of the table under 0, we find the triple 0RB, which means write the 0, move one cell to the right (R), then reenter state B. The same fate awaits a cell containing a 1. The machine basically ignores it and moves again one cell to the right, remaining in state B all the way to the right-hand cell with an X in it.

When the machine examines this cell, according to the table, it does a 0RC, so to speak. It replaces the X by a 0, moves one cell to the right, then enters state C. Readers hardly need me after this. In state C you will see that no matter what symbol may be found in the next cell, the machine writes an X, placing a new end marker, in effect. After that the machine enters a new state, D, which is the left-moving equivalent of state B. In this state the machine leaves each symbol as it is, moving left all the while, until it encounters an X. When this happens it enters its final, or halting state, E.

Another way to represent this machine's "brain" is by use of a diagram that merely echoes the table in graphic form. There each state is represented by a circle, and the transitions are represented by arrows. You may follow the progress of the machine from state to state more easily by looking at this diagram. The labels on the arrows are merely abbreviated versions of the triples in the table. The first symbol, near the root of an arrow, represents the symbol appearing in the cell being examined. The next symbol along the arrow represents the move to be made by the machine (left or right) along the tape. The arrow leads directly to the state that the machine enters next, whether it's the same state or a new one. The last state, E, has no arrows leaving it.

When a Turing machine has done whatever it is supposed to do, it halts. However, mere possession of a halting state is no guarantee that a given Turing machine will halt on a particular tape. For example, the doubling machine introduced above, if given a tape with only one X on it, will never halt, but continue moving to the right forever. One can object that such nonhalting behavior is trivial in the sense that we had to doctor the tape to prevent the machine from halting, but less trivial examples abound.

It might seem, from the utter simplicity of Turing machines, that they would not be capable of much. But it is not difficult to construct Turing machines that add or multiply numbers, that manipulate characters in the manner of a word processor, and a great deal else, besides. Turing machines can do anything! Well, anything that a computer can do, anyway. This was not immediately obvious in the 1930s, when Turing first formulated his machines. In fact, there were no digital computers at all at the time. That is why anything a Turing machine can do is called "Turing computable."

In the context of all possible Turing machines, we find ourselves facing not only more or less sensible machines that carry out sensible computations such as arithmetic but also all sorts of strange machines that carry out perfectly obscure, often nonsensical computations. Would you like to design your own Turing machine? The exercise is trivial. Just set up a state-transition table, as I have done on page 170, and fill it in with arbitrary symbols of the appropriate type.

I would be very surprised if the machine I have just "designed" did anything even mildly comprehensible. The point is that the theory toward which Turing was working had to include all possible Turing machines because that class and only that class embraced, in an inclusive mathematical sense, exactly what he meant by a computation.

In truth, each Turing machine is more like a program than a computer. Like the number doubler, it has a single mission in life and can carry out only that mission. For Turing, working on the theory of computability in

	O	1	X	#
A	# R A	O L A	1 N B	X R B
B	O N C	# R A	X R D	O N B
C	1 N B	# N B	1 L A	O R B
D	1 N D	O N A	O R D	O R C

An obscure (and probably useless) Turing machine

the early 1930s, it seemed that the machines he conceptualized nicely captured the idea of what it meant—to Turing, at least—to compute. What he needed, however, was a conceptual framework in which the myriad of machines might exist, so to speak.

The universal Turing machine, denoted by a U, provided this framework. Like the machines just discussed, the universal machine has a fixed state-transition table, but it has two tapes, as shown on page 171.

The upper tape contains a description of a particular machine, M, essentially its state-transition table written out as one long string of symbols. The lower tape is a duplicate of M's tape. The universal machine U proceeds by examining the symbol currently being read on M's tape. Then it looks up the appropriate triple on the tape containing the description of M in order to find out what M would do under those circumstances. It then writes the same symbol on M's tape that M would write, moves to the same next cell that M would move to, and keeps track of M's next state. By repeating this cycle of operations endlessly, the universal machine simulates perfectly the action of M on its tape. The fact that it takes a very long time to carry out this mission is meaningless in the theory of computation. The only sine qua non is that a finite number of steps be involved in any computation.

With the universal machine U in hand, Turing was ready for the next step. He was after big game, after all, to address the *Entscheidungsproblem* in this context. In particular, Turing wondered about what he called the

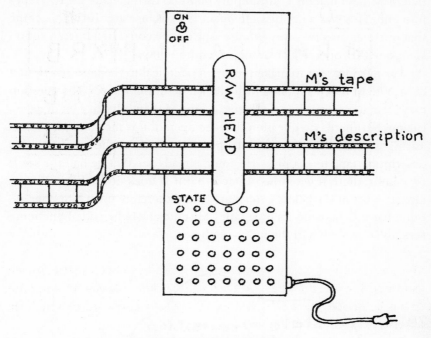

The universal Turing machine

"halting problem." Was there a general procedure for deciding, given an arbitrary Turing machine M and an arbitrary tape T for it, whether M would ever halt on T? The procedure would examine both M and T and, after a finite number of steps, report either "yes" or "no." Decision problems always have yes-or-no answers.

Turing solved this problem with a relatively simple construction that showed that the *Entscheidungsproblem* (for Turing machines) had no solution. His construction is so simple that the ultimate sounding board of all inquiry, the person in the street, could decide the matter for himself or herself.

He supposed that he was already in possession of a Turing machine D that can solve this particular decision problem. D would be like the universal machine in that it operated on a tape description of an arbitrary machine M, along with a copy, T, of M's tape. D has two halting states. If it finishes its computation in the state we will call Y, it means that M eventually halts on T. If D ends up in state N, however, it means that M will not halt on T. To simplify matters, Turing made the two tapes into one. This was a permissible operation because any Turing machine that operates on a two-way infinite tape can be converted into an equivalent one that operates on a one-way infinite tape.

Turing's idea was to make the tape T identical to the tape description of M. Will M halt if supplied with its own description tape? The question may seem nonsensical, as if we were asking M to indulge in a bit of navelgazing, but there's no question that the operation is meaningful. Worse, Turing wondered what would happen if D were supplied with its own description tape to examine. He could already see that he had the seeds of a contradiction when he noticed that if D does not halt on its own tape, it must nevertheless enter state N and therefore must halt after all. But what if D halts on its own tape? Mischievously he added one more state to D, as shown in the following figure.

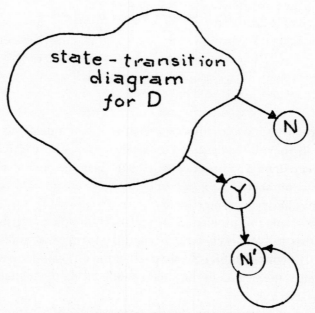

An indecisive Turing machine

He changed Y to a nonhalting state by adding a new transition arrow that goes to a new state, N′, in which, no matter what symbol D is scanning, it writes the same symbol and moves one cell to the right. This means that if D is supposed to halt on its own description tape, then it doesn't halt at all. To summarize, D halts on its own description tape if and only if it doesn't halt on its own description tape. The contradiction is firm and fully rounded, with no way out. D could not have existed in the first place, and the halting problem is unsolvable.

This was one of the first decision problems shown to be undecidable. Others followed in quick order.

THE MARTIAN DICTIONARY

The mid-1930s were remarkable years both in the early history of mathematical logic and in computing. Although digital computers had yet to be developed, their seeds were taking root in the fertile imaginations not only of Turing but of other gifted scientists as well. Just prior to the publication of Turing's influential paper "On Computable Numbers with an Application to the *Entscheidungsproblem*," the American logician Alonzo Church published a description of the lambda calculus, a system of generating strings of symbols that appeared, at the time, to have little or nothing to do with Turing's concept of computability. Another American logician, Stephen Kleene, published an account of recursive function theory in the same year. Who would suppose that recursive functions, defined by equations that use their output values for further inputs, had anything to do with the other two concepts?

Other systems were on the way. As early as 1914, the Norwegian mathematician Axel Thue, posed a problem involving the manipulation of symbols that would one day be recognized as the source of yet another alternative system. In 1947 the American logician Emil Post, independently but simultaneously with the Russian mathematician A. A. Markov, analyzed Thue's system, recognized it as equivalent to the other three, and found another unsolvable decision problem lurking in its heart.

Thue's original problem involved an alphabet and a dictionary. You could call it a Martian dictionary because it involved not words of English, but arbitrary words that look as though they came from Mars. Here's an example of the kind of problem Thue studied. It involves "translating" one curious-looking sentence (with no blanks) into another by substituting words within the sentence: RTCXUPNTRX.

Here are the ingredients of a more or less typical problem of this type:

The dictionary:

WORD	"DEFINITION"
PN	XXR
NTR	CU
XUP	CXXP
TCX	RNC
RX	NP

The problem: Transform the sentence RTCXUPNTRX into RRNCUXXRTNP.

The rules: To transform the first sentence into the second one, you must find a sequence of substitutions that does the job. Each substitution involves spotting, within the current sentence, one of the words in the left-hand column of the dictionary. You may then replace that word, within the sentence, by its opposite number in the dictionary.

The solution: In each sentence of the following sequence I have underlined the string being replaced.

R<u>TCX</u>UPNTRX   RRNCUPNT<u>RX</u>   RRNCU<u>PN</u>TNP   RRNCUXXRTNP

This particular version of Thue's word problem might seem pretty easy, but such problems can get really nasty, not to mention unsolvable. Actually, it's quite easy to pose an unsolvable word problem of this kind. For example, if I asked you to solve

TRCXUPNTRX → RXTNPUXCRT,

you would undoubtedly fail. This is not what I mean by a problem that is unsolvable by a computer, however.

What Thue asked for was a procedure or algorithm that would settle any and all examples of the word problem, one way or another, in a finite amount of time. If the problem has a solution, this procedure should find it. If the problem has no solution, the procedure should also discover that. In fact, the procedure does not even have to produce a solution, it merely has to answer "yes" or "no." Either a solution exists or it doesn't. Thue had unwittingly posed one of the first decision problems of the twentieth century.

I will demonstrate how Thue's system contains a computer of sorts, lurking inside it, by showing how it may simulate a Turing machine. It will turn out, after I have demonstrated this transformation, that asking for a solution to the resulting word problem is like asking whether a Turing machine will halt.

To set up this simulation, I must first identify certain sentences as Turing machine operations. So I will start with an unspecified Turing machine M that has

states $q_1, q_2, \ldots q_n$,

symbols 0, 1, X (and perhaps others),

and tape moves L, R, N.

I will begin by writing out all the symbols on the Turing machine's tape at some arbitrary point during its processing of the tape, as in the following example:

0 1 1 X 0 X 0 0 0 1 X 0 1 1 0.

∧

Suppose the machine is currently scanning the 1 (marked by a caret). To represent this situation, I will place the symbol q_i (the machine's current state) just to the left of the symbol currently being scanned:

0 1 1 X 0 X 0 0 0 q_i 1 X 0 1 1 0.

If you guessed that such strings were destined to become the "sentences" of our Thue system, you'd be right. What I need now are dictionary entries that reflect the Turing machine's operation. Suppose then that the operations that would apply when the machine is in state q_i and reading a 1 appear in the following row from M's state-transition table:

	0	1	X
state q_i:	q_k 0 N	q_j 0 L	q_l 1 R

If the machine is reading a 1 (center column) and happens to be in state q_i, it will enter state q_j, write a 0, and move one cell to the left. I can represent this situation by an appropriate dictionary entry, as follows:

0 q_i 1 q_j 0 0.

Going back to the odd-looking sentence above, I can now demonstrate the effect of this substitution:

0 1 1 X 0 X 0 0 0 q_i 1 X 0 1 1 0

becomes

0 1 1 X 0 X 0 0 q_j 0 0 X 0 1 1 0.

Although I have demonstrated the process of translating Turing machine operations into a Thue system with but one example, it should be clear to readers how the thing is managed in general. Simply replace each entry in the machine's state-transition table by a dictionary entry.

It now remains only to give the initial and final words, again by an example. To make things just a bit more concrete, I'll use the example of the number-doubling machine introduced earlier in this chapter (see page 166).

The initial and final tapes looked like this:

Initial: X 0 1 1 0 1 1 1 0 0 1 0 1 1 X

Final: X 0 1 1 0 1 1 1 0 0 1 0 1 1 0 X

The machine happened to have capital-letter state names, such as A through E, so I hope it does not throw any readers off if I use lowercase versions of these, instead of the q_i notation, here. Initially, the machine was in state *a* while scanning the leftmost X and, in the end, it was in state *e* while scanning the rightmost X. Thus our initial and final words must be:

a X 0 1 1 0 1 1 1 0 0 1 0 1 1 X

and

X 0 1 1 0 1 1 1 0 0 1 0 1 1 0 *e* X.

Just to illustrate the process, I will begin it. One of the words in the dictionary must correspond to what the Turing machine must do when in state *a* and reading an X. The instruction for this situation can be found as the following entry in the state table that we described earlier:

X R B.

In our present notation, this translates into the dictionary entry

a X 0   X *b* 0.

The lower-case state marker shifts one position to the right and changes from an *a* to a *b*. Applying this dictionary entry to the first word above, we obtain

X *b* 0 1 1 0 1 1 1 0 0 1 0 1 1 X.

After some fifteen more steps, the second word (above) will result:

X 0 1 1 0 1 1 1 0 0 1 1 0 *e* X.

This particular problem is easy in that there is literally no choice about what substitution to make at each stage of the solution process. But the fact that such deterministic Thue systems can simulate Turing machines means that they are just as powerful as computational schemes.

It immediately follows that the problem originally posed by Thue is unsolvable. Suppose that an algorithm existed that could decide, for each and every instance of the problem, whether it had a solution or not. Next, we translate an arbitrary Turing machine into its corresponding Thue system and let the algorithm loose on it. After a finite time it would tell us whether the final tape was achievable by the machine or not. If it answers "yes," the machine must have halted. If it answers "no," the machine could not have halted. But then it will have solved the halting problem for Turing machines, a thing we already know to be impossible. Ergo, the algorithm cannot exist, and the word problem for Thue systems is insoluble. The Martian dictionary doesn't always work!

IS THERE A WAY AROUND IT?

The short answer is: only if Church was wrong. The evidence that he was right, that anything remotely resembling some kind of system for automating thought turns out to be equivalent to all the other systems ever devised, is very powerful indeed. As far as algorithmic abilities go, humans and computers share this fundamental limitation.

It is nevertheless interesting to examine the roots of computation in these high adventures of mathematics and logic, as well as to probe the ultimate branches of future developments.

Turing was aware from the very beginning that his conceptual machine was capable not only of much greater sophistication but also of an actual physical realization. For example, the individual Turing machines were actually like programs in that each embodied a particular procedure or algorithm. The universal Turing machine, on the other hand, corresponded to the programmable computer. After all, it simply did what the individual machine told it to do, much as a programmable digital computer does what its program tells it to do.

During World War II Turing became familiar with digital principles when building special machines he called "bombes" that systematically worked through cipher intercepts from German U-boats. After the war he put these principles to work by developing one of the world's first computers at the Teddington Research facility by 1950, a small machine called the Pilot ACE. Like all early computers, this one performed the essential switching functions with vacuum tubes.

John von Neumann, another major contributor to the development of mathematical logic and metamathematics, made major design contributions to one of the first computers in the United States, the EDVAC, as well as a more advanced computer at the Princeton Institute for Advanced Study; the computer became operational in 1952. Both scientists recognized that the inherent power of these machines lay in the provision that they be programmable. Since the 1940s and 1950s, computers have become so all-pervasive that, should they all suffer some unknown malady and stop, our entire technical civilization might well grind to a halt.

Computers of the future, some beginning to take shape in the most rudimentary form, others yet gleams in someone's eye, include not only the current batch of "supercomputers" but also all-optical computers, DNA computers, and quantum computers. Will any of these machines, real or proposed, evade the strictures of Church's thesis?

The supercomputers of today are essentially very fast parallel machines. The parallelism means that they can carry out several instructions simultaneously, thus speeding them up. No matter how fast they get, however, they will never be able to perform more than a fixed, finite number of instructions at a time. If there's no algorithm for a particular problem, a machine can operate as fast as one likes, even doubling its speed every second; it will make no difference to the final outcome. No algorithm means no algorithm.

Optical computers will operate with light instead of electrons, but the limitations will remain in place. DNA computers promise some speedup by enabling strings of DNA to interact in a sort of "solution broth" in which a myriad of chance combinations will all be tried at once, en route to the solution of a problem. But those chance encounters between molecules will still be limited to a finite number, no matter how large the "tank." Thus DNA computers will fall under the same gloomy cloud.

The idea behind quantum computers, still an untested, albeit promising technology, exploits a physical phenomenon known as quantum wave collapse (see chaper 3). In its most speculative form, such a computer will produce a wave function that corresponds to, or encodes, a problem worth solving. If one forces the wave function to manifest itself by measuring it in the appropriate way, a choice will be made among all the alternative forms that the function could take. The simplest example of such a collapse occurs when individual photons are directed at a pair of slits in some otherwise impervious material. Each slit is equipped with a detector. It has been shown that when the detectors are turned off, the photons pass through both slits simultaneously. But when the detectors are turned on, the photon must "make a choice" about which slit to pass through. This is what is meant by the "collapse" of the photon's wave function. If further conditions can be placed on a wave function, conditions that correspond to the strictures of a particular problem to be solved, it may be that the wave will collapse in a manner that provides a solution to the problem.

It may even be that the two-slit experiment itself, or some very sophisticated version of it, can be adapted to solving decision problems: if the photon passes through slit A, the Turing machine will halt; if the photon passes through slit B, the Turing machine will not halt. This rosy scenario avoids any and all details (wherein the Devil waits) to encumber the scheme. However, troubles of a specific kind also may plague such devices. In the next chapter, which is devoted not to problems that can never be solved, but to those that merely take the age of the universe to solve, I will return to quantum computers for a closer look.

POSTSCRIPT

For readers who enjoy hand calculations that takes hours, even days, for programmers who enjoy the idea of driving computers crazy, for people who like numbers so monstrously huge they cannot even be vaguely conceived, there is Ackermann's function. This function, called A below, grows so fast in its values $A(1)$, $A(2)$, $A(3)$, . . . that it is *almost* not even recursive (i.e., computable). In any practical sense, it is not computable at all.

Here are four recursive equations that define Ackermann's function, denoted by the symbol A with a single variable within its parentheses. The function $S(n)$ denotes the successor of the integer n, namely $n + 1$. All computations start at m = 0:

$$A(0, n) = S(n)$$
$$A(S(m), 0) = A(m, 1)$$
$$A(S(m) \cdot S(n)) = A(m, A(S(m), n))$$
$$A(m) = A(m, m).$$

This obscure-looking set of formulas lists permissible substitutions in a process of computation. Starting with m = 0, for example, the equations become

$$A(0, n) = S(n)$$
$$A(S(0), 0) = A(0, 1)$$
$$A(S(0) \cdot S(n)) = A(0, A(S(0), n))$$
$$A(0) = A(0, 0).$$

What is the value of $A(0)$? It must be $A(0, 0)$, according to the last formula. What is $A(0, 0)$? If we put $n = 0$ in the first equation, we get

$$A(0, 0) = S(0).$$

The successor (S) of 0 is obviously 1, so we immediately have $A(0) = 1$. The next value, $A(1)$, of Ackermann's function is 3, and the value $A(2)$ after that is 7. It takes only a few minutes to work these out.

The fourth value takes a couple of hours and results in the value 61. Then the function really takes off, so to speak. It's not clear if any computer has enough digits to compute $A(4)$, and $A(5)$ may have more digits than there are fundamental particles in the universe! So forget $A(5)$ and everything beyond that. If you ever feel the need to exaggerate a claim, you can say, "Government spending is going up faster than Ackermann's function."

· 8 ·

The Big-O Bottleneck

Intractable Problems

> THERE ARE SOME MATHEMATICAL PROBLEMS
> THAT COMPUTERS CAN SOLVE ONLY BY TAKING
> AN EXPONENTIAL AMOUNT OF TIME.

EXPONENTIAL GROWTH IS EXPLOSIVE, to say the least. A computer that takes an exponential amount of time to solve a problem is also caught in an explosive situation because the amount of time it may take to solve the problem "blows up." As the computer tackles larger and larger instances of the problem, the amount of time it takes to solve them increases exponentially. It makes no difference how fast the computer is, exponential is exponential. Surprisingly small instances of the problem may take the lifetime of the universe to solve.

Problems that, for one reason or another, have no quick solution are called intractable. The amount of computation it takes to solve them never fails to remind me of a bottleneck. I call it the big-O bottleneck because "big O" is the notation for orders of magnitude. When numbers get really large, only their orders of magnitude may count.

The traveling-salesman problem is perhaps the best-known example of a seemingly intractable problem: A salesman travels a road (or air) network, visiting city after city to sell his goods. To keep profit high, he naturally wants to keep expenses as low as possible. This means keeping his mileage to a minimum. Therefore he must travel the network so he visits each city exactly once and so the total distance traveled is a minimum.

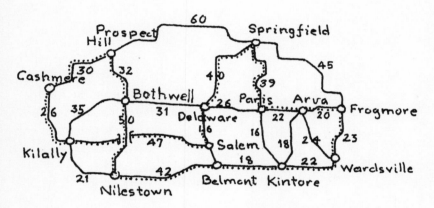

Network for traveling salesman

If, for example, he takes the route shown in the figure, his total distance traveled is 426 miles. But there is another route that involves just 422 miles, a somewhat shorter route. Can you find it?

Computers have been (and will continue to be) programmed to find the minimum-cost route for this problem. A program may take only 1 second to solve the problem shown here. But it may well take 2 seconds to solve a route involving just one more city, namely eleven cities. As for twelve cities, that could take as long as 4 seconds. Continuing this dreary arithmetic up to fifty cities, we find that it could take more than 3.5 million years for the program to find the minimum-length tour.

Why does it take so long to solve the traveling-salesman problem? There's no simple answer, but most of the solution methods devised to date must consider a great many alternative tours before settling on the one of minimum length. It will come as no surprise to the reader that the number of such tours increases exponentially with the number of cities. For example, for the map above there are 124 tours, but only 1 of minimum length. The more cities there are on a map, the more tours there are to consider, exponentially more.

Strangely, there are similar problems that do not require an exponential amount of time to solve. Suppose we take the same map as I have drawn above and ask for the shortest route joining two of the cities—say, Nilestown and Springfield. There are more than thirty routes from Nilestown to Springfield, and if I were to enlarge the map by one city, there might be twice as many. In other words, the number of alternative routes from one city to another may also grow exponentially with the number of cities, yet there is a method of finding the shortest route that does not require an exponential amount of time.

Both problems involve minimizing a distance. In the first problem, however, the tour's path must pass through every point in the network. In the second problem, there is no constraint on the path. It may pass through as few points as one likes, as long as it connects Nilestown and Springfield. Sometimes lifting a constraint makes a problem easier to solve.

ALGORITHMS AND PROGRAMS

In the mathematical sciences it is crucial for the objects of study to be well defined—that is, capable of exact expression. For example, I could formulate an operation on numbers called "enlarging." If I wanted to enlarge 10, I might well replace it by 20. I would not get very far in developing a theory based on this operation. The theory would depend crucially on just what I meant by "enlarging" a number. Do I mean doubling it? Do I mean adding 1 to it? Would I mean selecting a random number and adding it to the number in question?

An algorithm is a step-by-step procedure for solving a problem. To the extent that an algorithm is well defined, it can always be converted into a computer program by translating it into an appropriate computer language. For example, suppose L is a list of n numbers and that the kth member of the list is denoted by $L(k)$. Here is an algorithm for finding the largest number in L. I have labeled some of the lines or steps of this algorithm with numbers for easy reference.

1. Set *Largest* equal to $L(1)$

2. Set $k = 2$

3. If $L(k)$ > Largest,
    replace the value of *Largest* by $L(k)$.

4. If $k < n$,
    set k equal to $k + 1$ and return to step 3.
    Otherwise, output *Largest* and quit.

The italicized letters and names *Largest, L,* and k are variables. In a computing context, one may think of a variable as a kind of cubbyhole that will hold one number at a time. The variable *L* is actually a multiple variable that computer scientists call an array. The values of the array are indexed as $L(1)$, $L(2)$, $L(3)$, and so on.

Just for starters I will give the following values to *L* to illustrate the algorithm: $L(1) = 17$, $L(2) = 9$, $L(3) = 37$.

In step 1 the variable called *Largest* will receive the value 17 because that is the value of $L(1)$. In step 2 the algorithm sets the variable k to the value 2, and in step 3 the algorithm compares the sizes of $L(2)$ and *Largest*. Is $L(2)$ greater than *Largest*? Is 9 greater than 17? If 9 were, the algorithm would replace the value of *Largest* by 9—but it isn't larger, so *Largest* remains unchanged. Step 4 first compares k with n, the size of the array L. We're not there yet, so the algorithm increases the value of k by $k + 1$, turning the 2 into a 3, and then goes back to step 3. This time around, with $k = 3$, the algorithm compares $L(3)$ and *Largest*, finding that $L(3) = 37$ is indeed larger. Thus it replaces the old value of *Largest*— namely, 17—with the newer (and larger) value of 37.

Around and around the algorithm goes, in a loop, replacing the value of *Largest* by any number in the list *L* that happens to be larger yet. In the end, k is no longer less than n, and the very last line of the algorithm swings into play by outputting the last value of *Largest*. This will, of course, be the largest number on the list.

An algorithm is a peculiar thing. You will notice that it specifies actions but does not actually carry them out. Action is the province of the computer—or human being. Algorithms are programs in waiting, if you like. They can be executed by human beings (as I just did) or by computers as long as there is a human being around to convert the algorithm into a working computer program. In fact, algorithms are really vehicles of communication expressed in an informal, flexible language that is just precise enough to capture the essential elements of a method without having to go into any details. It is frequently the language of programmers.

The algorithm I have just introduced may be analyzed to find out how long it takes to find the largest number in a list of n numbers. In arriving at the time it takes, we worry only about how long it takes in the *worst case*. In other words, over all instances of a given length, how busy can we make this algorithm so that it takes the longest possible time to arrive at an answer?

To find out, we assume that each step of the algorithm takes one unit of time to execute. We may keep score by reproducing the algorithm and writing a dot beside each step as it is performed. To save a little time, I will simply assert that the worst case this algorithm will ever have to face occurs when the numbers in the list increase from first to last. Thus, for example, the list $L = (1, 2, 3, 4, 5, 6)$ will result in the following time score for the algorithm:

$$k = 2 \ \ 3 \ \ 4 \ \ 5 \ \ 6$$

1. Set *Largest* equal to $L(1)$ ·

2. Set $k = 2$ ·

3. If $L(k) > Largest$, · · · · ·

    replace the value of *Largest* by $L(k)$. · · · · ·

4. If $k < n$, · · · · ·

    set k equal to $k + 1$ and return to step 3. · · · · ·

    Otherwise, output *Largest* and quit. · · · ·

    ·

To evaluate the score, simply count the dots. Each dot represents a step in the algorithm for a particular value of k. Under the k value of 2 there are six dots, for example, because the first six steps are executed. For each of the remaining values of k, the algorithm executes four steps each. The total number of dots is twenty-two, so the algorithm takes twenty-two steps to find the largest in a list of six numbers. This is the worst case, of course. If the order of the numbers is reversed, the algorithm would have five steps for $k = 2$ and three steps each time after that for a total of seventeen steps.

To find out how well the algorithm would do in general, we assume the list has n numbers in it and ask how many steps it would take, in the worst case, for the algorithm to come up with an answer. The answer comes by generalizing what we already know: it will take six steps for $k = 2$ and four steps for each value of k from 3 to n. The total number of steps will therefore be $6 + 4(n - 2)$, or $4n - 2$.

This particular algorithm therefore takes $4n - 2$ steps to solve the maximum-number problem in the worst case and, since the expression for the time taken is linear, we say that the algorithm takes linear time to solve the problem. It turns out that when we are faced with problems that take an exponential amount of time, the coefficient 4 and the additive constant -2 amount to pretty small change, so we ignore them. Only the factor n is important.

Can the maximum-integer problem be solved any faster than in linear time? We can make the argument that no algorithm could possibly discover the largest integer without examining all n numbers of the list (in the worst case), so a linear-time algorithm is optimal. There is no algorithm, for example, that takes only logarithmic time, nor could there be.

Strangely enough, if I change the problem slightly, there is an algorithm to solve it in logarithmic time. In the new problem, I want to see if a particular number m happens to be in the list. The list, it turns out, has already been sorted into increasing order. How long would it take, in the worst case, to discover whether m is on the list? The answer involves a ubiquitous technique in computer science: binary search.

There is an amusing anecdote that illustrates the famous binary searching algorithm: How do you trap a lion in the desert? First you build a fence across the middle of the desert. The lion will end up on one side of the fence or the other. Next, you build a second fence across the half of the desert that is presently inhabited by the lion. Again, it must be in one or the other of the new halves (now constituting one-quarter of the desert). The picture should now be clear. At each iteration of the basic lion-trapping algorithm, one cuts the range in half. Before long, the lion is inside a rather small pen. You have trapped it with the help of applied computer science.

The binary search algorithm looks for a given number m in a list or array L of n numbers arranged as above. The algorithm first examines the number in the middle of the array. If m is less than this number, the algorithm "knows" it must be in the first half of L; otherwise, it lies in the second half. Having determined which half holds the number, the algorithm next searches that half by exactly the same method. At each iteration of the basic scheme, the range to be searched narrows by half.

If you think that continued doubling leads to rapid growth, consider how continued halving leads to rapid shrinkage. If k doublings lead to a number in the order of $2k$, then k halvings lead to a number such as $\log(k)$, and this is precisely the worst-case time taken by the binary searching algorithm.

There are algorithms of all possible complexities. The following figure displays just a few of them. The lowest curve shows how the computation time grows for an algorithm of logarithmic complexity. The straight line illustrates linear complexity. The lower upward-opening

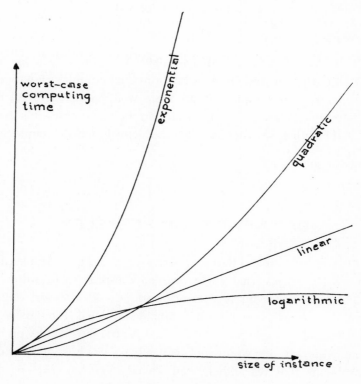

Plots of various complexities

curve represents quadratic complexity. The highest curve demonstrates how exponential complexity runs "screaming off into infinity," as a colleague once put it.

Although the quadratic-time algorithm also looks like it will take a long time, given large enough examples of the problem it solves, it really doesn't amount to a hill of beans compared to how much time the exponential algorithm takes on instances of the same length. All problems that can be solved (in the worst case) by polynomial-time algorithms, even ones that take

$$n^{284}$$

steps to solve, are considered "tractable."

To symbolize our rather cavalier attitude to additive and multiplicative constants in the expressions for complexity of an algorithm, we use the "big O" notation. The "O" stands for "order of magnitude." Thus an algorithm that takes $17n^2$ steps or even $23917n^2$ steps, where n is the length of the instance, is said to have complexity

$$O(n^2).$$

Other complexities are treated similarly. For example, linear complexity would be written $O(n)$ and exponential complexity would be written $O(2^n)$. Some complexities are more, well, "complex." If an algorithm takes on the order of $n\log n$ steps (in the worst case) to solve an instance of length n, we would say that the algorithm has a complexity of

$$O(n\log n).$$

BAD NEWS FROM BERKELEY

In 1974, Stephen Cook, a Berkeley graduate student, made a startling discovery. His thesis, that some problems appear to be inherently intractable, shook the computer science world. Cook's discovery centered on a well-known problem in propositional logic called "the satisfiability problem." Cook proved that if one could solve the satisfiability problem reasonably quickly, one would automatically solve a host of other difficult problems that researchers had slaved over for years—getting nowhere. Cook's result implied that the satisfiability problem must be very hard to solve indeed. More importantly perhaps, his theorem illuminated an old theme of mathematics, the search for equivalence among problems.

The problems that formed the focus of Cook's research were all decision problems (see chapter 7). They all had yes-or-no answers. One can always convert a more regular problem such as the traveling-salesman problem into one that has yes-or-no answers by posing it in a slightly different way. Instead of asking "What is the minimum-length tour of this particular network of roads?" we can ask "Does this network contain a tour of length less than 25?" If we can always get an answer quickly to the latter sort of question, we can, by playing twenty questions, obtain an answer to the original problem. Does it have a tour of length less than 48? Yes? Does it have a tour of length less than 24? No? Does it have a tour of length less than 36? No? And so on. Astute readers will see the binary search technique at work here.

Given an instance of the satisfiability problem, a solution algorithm would have to answer the question "Does this expression have a satisfying assignment?" In other words, it would have to find values for the variable in the expression that made it true. Here is an instance with four logical variables, $w, x, y,$ and z:

$$(w + x' + y' + z) \cdot (x + y') \cdot (w' + y + z).$$

The expression consists of a logical product ($\cdot$) of logical sums (+) or clauses, as I shall call them. The ultimate constituents of such an expression are called "literals." These are the logical variables, either as they stand, x, or negated, x'.

Depending on the values (0 or 1) of the logical variables, the expression itself will have a logical value. In this context, "true" is symbolized by "1," while "false" is indicated by "0." A satisfying assignment would be a collection of values for the variables w through z that produced a value of 1 for the expression as a whole.

For example, the solution

$w = 1$

$x = 1$

$y = 0$

$z = 1$

satisfies the expression because every sum such as $w' + y + z$ takes the
value 1 after the dust settles: $(0 + 0 + 1) = 1$.

The expression itself becomes

$$(1 + 0 + 1 + 1) \cdot (1 + 1) \cdot (0 + 0 + 1) = 1 \cdot 1 \cdot 1$$
$$= 1 \text{ (true)}.$$

The sum of two or more literals follows the logic of "or." Under this
logic $0 + 0 = 0$, $0 + 1 = 1$, and $1 + 1 = 1$. The first clause, $1 + 0 + 1 + 1$,
therefore equals 1. In fact, all three clauses equal 1 and their product
must also equal 1, or true.

The product of three true clauses is again true. Thus the given assignment satisfies the given expression, and the given values do, indeed, con-
stitute a solution. This particular instance has many solutions, but some
instances have just one solution, while others have none at all. In general, the satisfiability problem is very difficult to solve.

The length, n, of the expression may be taken as the number of literals in it, in this case 9.

How long does it take to solve the satisfiability problem? As the number n of literals increases, is there an algorithm that takes only a polynomial (in n) number of steps to solve it? Probably not, according to the
freshly minted Ph.D. Cook had proved that if such an algorithm existed,
it could be transformed into a solution for any problem in an extremely
broad class that he called "NP," short for nondeterministic polynomial-
time problems. I will discuss what "nondeterministic" means in a
moment.

Suffice it to say that the class NP contained not only the problems
for which workers had found polynomial-time algorithms, but also virtually all of the problems (like the traveling-salesman problem) that had
given hundreds of mathematicians and computer scientists so much
trouble over the years.

Cook had discovered a transformation from the entire class NP into
a single problem, satisfiability. The transformation was generic; it could
be applied with minor modifications to any problem in NP. In short,
given any problem in NP and any instance of that problem, Cook's transformation would convert that instance into an instance of the satisfiability problem. Moreover, it produced that instance in polynomial time. The
situation can be sketched with a simple diagram:

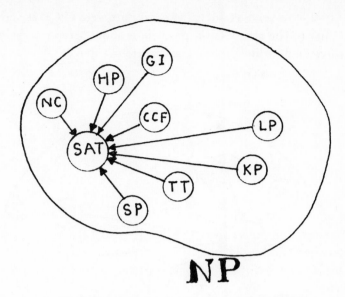

Transformation from NP to the satisfiability problem

The arrows show the action of the transformation on a great variety of problems, each with its own acronym. I will list them without explaining what they all mean HP (Hamiltonian path problem), SP (shortest-path problem), LP (linear programming problem), GI (graph-isomorphism problem), CCF (computer-circuit-fault-detection problem), NC (*nxn* checkers), KP (knapsack problem), TT (timetable design), and, of course, SAT (the satisfiability problem). Such a list only scratches the surface of the thousands of problems that lie in the class NP. As you can infer from some of the names, they include not only games and puzzles, but also problems of great importance for a society driven by technology.

If the humble reader should discover a polynomial-time algorithm for the satisfiability problem, he or she would, thanks to Cook's theorem, earn undying fame as the solver of all the problems in NP! Perhaps you don't have the experience necessary to discover such an algorithm. Perhaps you don't have the smarts. Perhaps you don't have the ambition. Yet, for some of the problems in NP, people with all these qualities in abundance have failed miserably. Chances are, some of these problems are inherently intractable. Let me show you what I mean.

Cook also found a polynomial-time transformation from satisfiability to other problems such as the vertex-cover problem (VC), as shown in the next figure. The vertex-cover problem is about "graphs," systems of points connected by lines, a problem mentioned above.

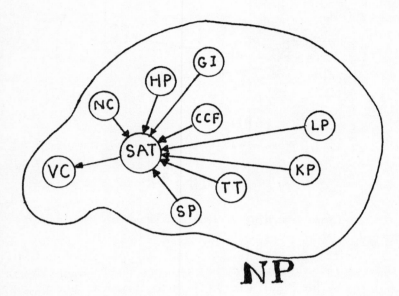

Transformation from SAT to VC

The vertex-cover problem turned out to have the same magic property as the satisfiability problem: find a polynomial-time solution for this problem and you will have found a polynomial-time solution for the whole class NP. The name of this magical property is "NP completeness." Within a few years of Cook's theorem appearing, researchers had found a chain of polynomial-time transformations extending out from these humble beginnings to more than a thousand different problems. And none of the problems thus shown to be NP complete appeared to have a polynomial-time solution algorithm. Moreover, none of the problems that appeared in the chain was among those that, like the shortest-path problem, were known to have polynomial-time algorithms.

There is a well-known cartoon among complexity theorists that illustrates the central significance of Cook's theorem and all the subsequent proofs of NP completeness. The poor researcher cannot find an efficient algorithm for a certain problem. He fears that he may have to confess to his boss, with no excuse for his failure, and the first frame of the cartoon shows him humbly apologizing. Of course, he would like to say that he failed because the task just wasn't possible, and the second frame of the cartoon shows him unapologetically telling the boss that the problem can't be solved. If his problem turns out to be NP complete, however, he can tell the boss that neither he nor a host of brilliant researchers can come up with an efficient algorithm. The legion of brilliant researchers has not worked on the problem of our poor scientist, but they have strained their brains to the uttermost on problems that are also NP complete. Had the hapless researcher solved his problem, he also would have solved theirs, and vice versa.

Before explaining the near-miraculous transformation of Cook's theorem, I will explain how one of the proofs of NP completeness works. Suppose I have a problem A, which I happen to know is NP complete. If I suspect that a second problem, B, also is NP complete, I might be able to find a transformation T from A into B. The transformation must have certain properties, however.

First, T must preserve truth values. If X is an instance of A, then T will produce an instance $T(X)$ of B. Let us call the latter instance Y. Such a transformation also would be expected to transform a solution of instance Y into a solution of instance X. The answer to the question associated with instance X will be "yes" if and only if the answer to the question associated with problem B is also "yes."

Second, the transformation must operate in polynomial time; there must be a polynomial $p(n)$ such that if the length of instance X is n, the length of instance Y can be no greater than $p(n)$.

Admittedly, all that is rather abstract, but let's press ahead to see what happens. Suppose I have also discovered an algorithm that solves problem B in polynomial time, the polynomial being q. Applied to an instance Y of problem B, the algorithm takes no more than $q(m)$ steps, where m is the length of an instance of B. The algorithm for problem B can now itself be transformed into an algorithm for problem A simply by first applying the transformation T to an arbitrary instance X of A, producing an instance Y of B. Next, I apply the algorithm for B to the instance X, discovering a solution (if there is one). If the transformation takes $p(n)$ steps, then the solution to the instance Y takes $q(p(n))$ steps.

What is the latter expression? It's a polynomial with another polynomial substituted into it. The result is also a polynomial, which is all we need to know. We have converted a polynomial-time algorithm for problem A into a polynomial-time algorithm for B. If I can solve problem A in polynomial time, I can do the same thing for problem B.

Problem B must therefore itself be NP complete, for if I now mate the transformation T with the generic transformation of Cook's theorem, I can transform any problem in the class NP into problem B, and I can do it in polynomial time. Moreover, any polynomial-time algorithm for problem B would become, in effect, a polynomial-time algorithm for every problem in NP.

I will now backtrack and give an example of a transformation from problem A (satisfiability) into problem B (the vertex-cover problem). In the satisfiability problem, every instance consists of a product of clauses. The vertex-cover problem (mentioned earlier) involves objects called graphs—namely, sets of points or vertices, some pairs of which are joined by lines or edges. Can I find a vertex cover for the following graph?

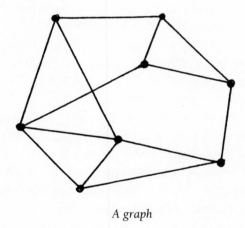

A graph

Such a cover would be a subset C of the points with the property that every line (or edge) of the graph has at least one of its vertices in C. Of course, if we put all the vertices of the graph into C, it will certainly be a cover. The real problem is to find the minimum number of vertices that C may contain in order to cover all the vertices of the graph. In the graph the minimum number is 5.

No one has ever found a polynomial-time algorithm for the vertex-cover problem. The possibility that no one ever will becomes very real, in light of the transformation I will now demonstrate. But how on Earth does one turn an instance X of the satisfiability problem into an instance Y of the vertex cover problem?

The transformation of any instance of the satisfiability problem into an instance of the vertex-cover problem will convert a logical expression such as the one we examined earlier into a graph. Here's how it's done. Replace each clause of the expression by as many vertices as there are literals in the clause, then join all the vertices in each clause set by edges. Next, replace each of the logical variables by two vertices joined by an edge, as in the following figure. Finally, join any literal in the upper part of the graph to a matching literal in the lower part. The figure below shows what happens when you apply this transformation to the instance of satisfiability that we examined earlier.

Here we see that the clause $(w + x' + y' + z)$ has become a cluster of four vertices all joined by edges. Clearly, any cover for the graph shown will have to include at least three of these vertices, so that all the edges are covered. One of the vertices—z in this case—is not in the cover. Therefore the edge that leads down to the lower part of the figure from z had better have its other end in the cover. It does.

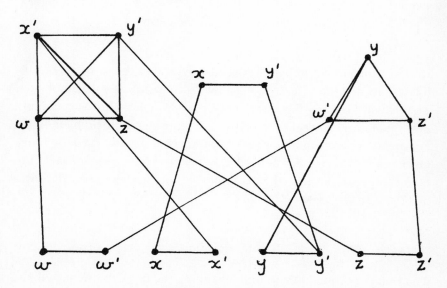

An instance of the VC problem

The same thing goes for the other two parts of the graph associated with the remaining two clauses. Each has all but one of its vertices in the cover, and the edge leading from the missing vertex to the lower part of the figure has its other vertex in the cover.

Do you see how, in the lower part of the figure, each horizontal edge has just one vertex in the cover? Could that represent a satisfying assignment? The answer is "yes." In fact, the recipe for the transformation goes like this: Start with any instance of the satisfiability problem and replace each variable x by an edge, (x, x'), as above. Next replace each clause by a complex that consists of the same number of vertices, each labeled with the name of the literal it represents. Finally, join each literal in the edges of the first step with the corresponding literal in any of the complexes. That completes the transformation.

The resulting graph must have at least m vertices in its cover, where m is the number of literals in the initial expression. If the corresponding question is "Does the graph have a cover with m vertices?" the answer is "yes" if and only if the corresponding logical expression has a satisfying assignment.

How long does the transformation take? Surely a polynomial number of steps. In the worst case, there is only one clause, and the transformation would have to create a graph complex with as many edges as there are pairs of variables, namely a quadratic number of them. The complexity of the transforming algorithm would therefore be

$$O(n^2).$$

The transformation just described is typical of the secondary transformations that produce the ever-growing list of NP-complete problems. The primary transformation discovered by Cook is, not surprisingly, more complicated because it focuses not on any specific problem in NP, but on all of them simultaneously. To understand Cook's generic transformation we must understand what a nondeterministic algorithm is. But before plunging ahead, the "complexity" of this chapter calls for a breathing spell.

REPRISE

The first conceptual key to understanding the big-O bottleneck is the notion of an algorithm. An algorithm consists of computational steps that lead to the solution of a problem. To be more precise, I should say "instance of a problem," since a problem such as the traveling-salesman problem can be considered as the collection of all possible instances of it—in this case, all possible networks to be traversed by traveling salesmen. The size of an instance is measured by some natural quantity associated with it, such as the number of cities in a network.

One can analyze an algorithm to discover its worst-case performance over all instances of a given size. The algorithm's complexity is just the total number of steps it takes to solve the worst-case instance of size n. Instead of exact-complexity figures such as $3n - 7$, $12n^2 - 3n + 25$, or $2^n - 12n$, we boil the complexity down to its essential size, using the big-O notation. The foregoing complexities then become $O(n)$, $O(n^2)$, and $O(2^n)$, respectively.

Stephen Cook showed that a particular problem, that of satisfying propositional expressions, was just as hard a problem to solve as any problem in a large class that I will define in the next section. If this difficulty is regarded as a disease, I have shown how the disease "spreads," with a single example. There is a way to transform the satisfiability problem into the vertex-cover problem for graphs so that (a) every instance of the satisfiability problem is translated into an instance of the vertex-cover problem, (b) any solution of the latter instance is translated into a solution of the original instance of the satisfiability problem, and (c) the transformation itself takes only a polynomial number of steps to achieve.

Because Cook's theorem transforms a potential infinity of problems into the satisfiability problem, however, the transformation in the proof is more complicated than most of the particular transformations from satisfiability "outward" into the vast crowd of problems that turn out to be just as difficult as satisfiability. We are about to find out just what that mysterious term "NP completeness" actually means.

NONDETERMINISTIC ALGORITHMS

The initials NP stand for "nondeterministic polynomial time." A problem in the class NP can be solved by a nondeterministic algorithm in polynomial time. The algorithms we have examined up to now were all deterministic. In other words, they proceeded to find solutions to instances of a problem by following a rigid recipe, the outcome of which is determined in advance. Not so with nondeterministic algorithms.

Faced with the instance of a problem, a nondeterministic algorithm simply guesses the yes/no answer and, if the answer is "yes," also guesses a solution. For example, the problem might be the traveling-salesman problem, and the question might be "Does the instance in hand have a tour of length less than 123?" The answer may well be "yes," in which case the algorithm not only says so, but also produces a tour of length 122 or less. By definition a nondeterministic algorithm is never wrong.

That's what "nondeterministic" means: guessing. How does polynomial time come into it? A nondeterministic algorithm also must check that the solution does, indeed, enable it to answer "yes" to the question. It enters a verification stage of processing during which it checks that (a) it has guessed a genuine tour and not some meandering route that passes through the same city twice or misses another city altogether, and (b) that the total length of its tour is less than 123 units. The checking process, the only measurable activity of the algorithm, must not take more than a polynomial number of steps.

Suppose that the map has seven cities, each labeled with a number, and that the tour guessed by the algorithm came in the form of a list of cities (in the order that the tour passes through them: 3, 5, 1, 7, 4, 2, 6, 3), the first city appearing at the end of the list. Taking n as 7, it seems clear that the algorithm would take in the order of n steps to check part (a). After all, it only has to scan the list once, ensuring that each city is connected to the previous one by a road. It then scans the list once more, consulting the table for the distance from each city in the list to the next one, then adding up all the distances. This also absorbs only a linear number of steps.

If the checking phase of the algorithm comes up with a total distance of 105, it has clearly found a tour of length less than 123, and it will answer the original question, "yes."

Clearly, nondeterministic algorithms are extremely powerful. Unfortunately, they exist only as theoretical constructs; yet that very power is what we may be demanding of an algorithm that could solve a problem such as the satisfiability problem in polynomial time.

THE GRAND GENERIC ALGORITHM

A problem in NP, as I explained earlier, has a polynomial-time, nondeterministic solution algorithm. We will treat one such problem—call it problem A—in the abstract. Enter the Turing machine.

Suppose we have a guessed solution for an instance of problem A. The algorithm that checks the solution can be recast as a Turing machine program (see chapter 6), and the instance can be represented by a string of symbols on the Turing machine's tape. Cook's grand generic transformation operates on the Turing machine program and its tape, converting both into a satisfiability problem.

The actual transformation is laden with details, but I can provide enough glimpses of its inner machinery that readers might at least say "I see." The glimpses, as such, will involve three different aspects of the Turing machine program: the next state, the symbol to be written on the tape, and which way to move the read/write head.

Suppose we are given just one "instruction" from a Turing machine program. It may be written as a quintuple of symbols:

$$(q_1, s_1, q_2, s_2, d).$$

This instruction says that if the Turing machine happens to be in state q_1 while scanning the symbol s_1 on its tape, it should next enter state q_2, replace the symbol s_1 by the symbol s_2 on the tape, then move the read/write head in the direction d, which might be to the left, the right, or remain stationary. Because time is of the essence in complexity studies, we will keep track of it through the variable t.

The three actions of the machine prescribed by this instruction involve entering a new state: writing a symbol and moving the read/write head. Each action will be represented by a logical expression. The variables that enter the expression will not be as simple as the ones I used earlier in the satisfiability problem. An intermediate step eases the transition.

If the machine is in state q_1 at time t and reads the symbol s_1 at time t, then at time $t + 1$ it will be in state q_2. I will rewrite this statement in logical form as follows:

$$Q(1, t) \cdot S(1, t) \cdot R(k, t) \rightarrow Q(2, t + 1).$$

There are three logical variables in this expression. The variable Q, subscripted in effect by number pairs such as $(1, t)$ represents the behavior of the Turing machine's states. If the machine is in state 1 at time t then $Q(1, t)$ must be true and not otherwise. Similarly, $S(1, t)$ is true if the machine happens to be reading symbol s_1 on its tape at time t, while $R(k, t)$ is true if the machine's read/write head is positioned over the kth square of the tape at time t.

The expression is written as an implication, as symbolized by the arrow. If $Q(1, t)$ and $S(1, t)$ and $R(k, t)$ are all true, then $Q(2, t + 1)$ also must be true. To cast the expression in the right logical form, I will get rid of the implication arrow using the fact that $A \rightarrow B$ can always be replaced by the logically equivalent expression $A' + B$—that is, "not-A or B." The expression above therefore becomes

$$[Q(1, t) \cdot S(1, t) \cdot R(k, t)]' + Q(2, t + 1)$$

or

$$Q'(1, t) + S'(1, t) + R'(k, t) + Q(2, t + 1),$$

which is clearly a clause of the sort that appears in the satisfiability problem. The literals look different because the variables have been indexed, as in the expression $(1, t)$.

Similarly, we may express the idea that if the machine is in state q_1 and reads symbol s_1, it will write a new symbol, s_2, as follows:

$$Q(1, t) \cdot S(1, t) \cdot R(k, t) \rightarrow S(2, t + 1).$$

Again, this becomes the sum

$$Q'(1, t) + S'(1, t) + R'(k, t) + S(2, t + 1).$$

Finally, the same thing is done for the third kind of action, moving the read/write head. Therefore the entire Turing machine program can be written as a lengthy product of such sums. No matter which instance of the problem we thus encode in the logical expression, the length of the encoded program depends mainly on the extent or range of the time variable t. How big will t get? This will be the time it takes the checking component of the nondeterministic algorithm to determine that a guess is correct. Because problem A came from the class NP, there is a polynomial $p(n)$ that acts as an upper limit on the time t that the algorithm takes to solve the instance of A. Here n is the length of the instance. This number also limits the number of tape squares that can be visited by the Turing machine that does the checking, so the total length of the clause system produced by Cook's generic transformation cannot exceed the product of $p(n)$ with itself, namely on the order of $p^2(n)$ steps.

Suffice it to say that the instance of satisfiability that is produced by Cook's transformation when applied to an instance of problem A will have a satisfying assignment if and only if the instance of problem A has a solution for the question being asked.

The significance of Cook's theorem goes beyond the question of whether efficient algorithms can be found for this problem or that one. The theorem has given us a glimpse of a new kind of mathematics, in my opinion, one that we may see more of in the new millennium. It has often happened in the history of mathematics that two problems originally thought of as quite different turn out to be the same or at least equivalent. There is a way of looking at the first problem that makes you realize that you're really looking at a disguised version of the second problem. The "way of looking" at it may amount to a transformation of one problem into another, and in this sense, Cook's theorem points the way.

IS THERE A WAY AROUND IT?

A number of proposals for new ways to compute have been publicized in recent years. DNA computing and quantum computing, in particular, have been touted as the solutions to our computing bottlenecks.

In DNA computing, a problem is encoded in a strand of DNA, and potential solutions are allowed to proliferate within a chemical broth seething with genetic molecules. Such computers are thought by some researchers to have the potential to squeeze through the big-O bottleneck because so many potential solutions can be considered simultaneously, in effect, that discovering a solution that works should take next to no time at all. For example, one strand of DNA might represent an instance of the traveling-salesman problem, while others might represent potential minimum-length tours. All the strands are replicated in the thousands or even millions and, as the chances of a solution strand matching a problem strand climb to near certainty, the presence of a solution is detected chemically.

In essence, the DNA is a vast parallel computer in which many processes can go on concurrently. It can be represented by a more traditional parallel computer in which there are k processors that, when suitably programmed, can operate in parallel. What effect does the presence of k processors have? It reduces the complexity of an algorithm by a factor of k. Unfortunately, k is fixed for the machine in question, and if the algorithm running on our hypothetical machine runs in time that is bounded by a polynomial $p(n)$, the net effect is to divide the polynomial through by k. As far as complexity theory goes, this changes nothing. As you may recall, all constant factors and terms were ignored in evolving the big-O notation. If you divide a quadratic polynomial by 1,736, you still get a quadratic polynomial when the dust settles.

The fact that each potential solution molecule can contact only a limited number of strands of problem DNA in the computer broth means that the degree of parallelism is limited for this kind of computer, just as it is for a standard parallel processor. This remains true even if there are billions of strands of each potential solution and problem instance. After all, a broth of fixed volume will contain a finite number of molecules, no matter how many. The number of potential encounters is therefore still limited, on average, by the number of copies of potential solution strands.

If successful, DNA computers may well speed up the search for solutions in smaller examples of the traveling-salesman problem, for instance, but it will yield this advantage only to the extent of extending the range of our computation to a handful of additional cities. Exponential is exponential!

The prospect for quantum computing is harder to evaluate. For one thing, there are several proposals extant at this time. The basic idea involves the infamous "collapse of the wave function" (see chapter 7). If a complex quantum system is in a premeasurement state, it may be thought of as the superimposition of a myriad of states. For example, 1,000 photons on their way to a polarization detector may be viewed as the superimposition of

$$2^{1,000}$$

states of polarization, two for each photon. Each combination of states has an exceedingly tiny but positive probability of manifesting.

The idea is to set up the measurement apparatus in a way that encourages the appearance of certain combinations of states while excluding others. While it may be too early to tell how effective this idea will be, is it too early to ask whether such quantum devices will amount to analog computers?

An analog computer operates by a principle of analogy. For example, suppose I want to solve the so-called minimum Steiner tree problem: Given a finite number of points in the plane, find a tree structure that (a) contains all the points and (b) has minimum overall length. The following figure illustrates a solution to the problem obtained by means of an analog "computer" I once built. It consists of two parallel sheets of Plexiglas with rubber-tipped pegs between the sheets. The arrangement of pegs represents the given problem instance.

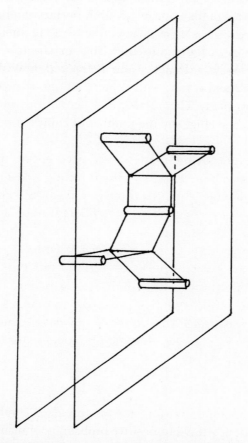

A soap solution to the minimum Steiner tree problem

Simply dip the apparatus in a soap solution and draw it out vertically. A soap film will automatically appear, and its configuration will resemble a minimum Steiner tree. In many cases it will actually be a minimum Steiner tree but, alas, in other cases the minimum arrived at by the soap film will be a local minimum and not a global one. In other words, given that that particular way of joining up the points (pegs) by lines (sheets of soap film) may be minimal for that particular geometry, there may be another geometry altogether that not only has a shorter overall length but also is not achievable by dipping the apparatus, no matter at which angle it is withdrawn from the bath.

Should it surprise us that the minimum Steiner tree problem happens to be NP complete?

Unless and until we stumble upon a completely new paradigm for what it means to compute, many of our more important problems will be solved only by forcing them through the big-O bottleneck. The bottleneck, like the other barriers to knowledge or action in this book, challenges us to the very limits of our competence. Will we ever break through? Don't bet on it.

· R E F E R E N C E S ·

CHAPTER 1: **The Energy Drain: Impossible Machines**

Atkins, P. W. *The Second Law.* New York: Scientific American Library, 1984.

Dircks, Henry. *Perpetuum Mobile: Search for Self-Motive Power during the 17th, 18th, and 19th Centuries.* London: E. & F. Spon, 1861.

Halliday, David, and Robert Resnick. *Fundamentals of Physics.* New York: John Wiley & Sons, 1988.

CHAPTER 2: **The Cosmic Limit: Unreachable Speeds**

Bergmann, Peter Gabriel. *Introduction to the Theory of Relativity.* New York: Dover, 1976.

Clark, Ronald W. *Einstein: The Life and Times.* New York: World, 1971.

CHAPTER 3: **The Quantum Curtain: Unknowable Particles**

Herbert, Nick. *Quantum Reality: Beyond the New Physics.* Garden City, N.Y.: Doubleday, 1985.

Hughes, R. I. G. *The Structure and Interpretation of Quantum Mechanics.* Cambridge, Mass.: Harvard University Press, 1989.

Rae, Alastair. *Quantum Physics: Illusion or Reality?* Cambridge, Eng.: Cambridge University Press, 1986.

Stapp, Henry P. *Mind, Matter, and Quantum Mechanics.* Berlin: Springer-Verlag. 1993.

CHAPTER 4: **The Edge of Chaos: Unpredictable Systems**

Cole, Franklyn W. *Introduction to Meteorology.* New York: John Wiley & Sons, 1980.

Gleick, James. *Chaos: Making a New Science.* New York: Viking Penguin, 1987.

Peitgen, Heinz-Otto, and Dietmar Saupe. *The Science of Fractal Images.* New York: Spinger-Verlag, 1988.

CHAPTER 5: **The Circular Crypt: Unconstructable Figures**

Beggren, Lennart, Jonathan Borwein, and Peter Borwein. *Pi: A Source Book,* 2nd ed. New York: Springer-Verlag, 2000.

Hellman, Hal. *Great Feuds in Science.* New York: John Wiley & Sons, 1998.

Jesseph, Douglas M. *Squaring the Circle: The War between Hobbes and Wallis.* Chicago: University of Chicago Press, 1999.

Niven, Ivan. *Numbers: Rational and Irrational.* New York: Random House, 1961.

CHAPTER 6: **The Chains of Reason: Unprovable Therorems**

Dawson, John W. Jr. *Logical Dilemmas: The Life and Work of Kurt Gödel.* Wellesley, Mass.: A. K. Peters, 1997.

Dewdney, A. K. *The New Turing Omnibus.* New York: W. H. Freeman, 1993.

Newman, James R., ed. *The World of Mathematics,* vol. 3. New York: Simon & Schuster, 1956.

CHAPTER 7: The Computer Treadmill: Impossible Programs

Dewdney, A. K. *The New Turing Omnibus*. New York: W. H. Freeman, Computer Science Press, 1993.

Hodges, Andrew. *Alan Turing: The Enigma*. London: Burnett Books, 1983.

Kleene, S. C. *Introduction to Metamathematics*. Groningen, The Netherlands: Wolters-Noordhoff and North Holland Publishing, 1971.

Minsky, Marvin L. *Computation: Finite and Infinite Machines*. Englewood Cliffs, N.J.: Prentice-Hall, 1967.

CHAPTER 8: The Big-O Bottleneck: Intractable Problems

Garey, Michael R., and David S. Johnson. *Computers and Intractability: A Guide to the Theory of NP Completeness*. New York: W. H. Freeman, 1979.

·FURTHER READING·

Some of the older books listed below are real chestnuts and are well worth collecting, if they can be found in a used bookstore or online.

Banchoff, Thomas F. *Beyond the Third Dimension: Geometry, Computer Graphics, and Higher Dimensions*. New York: Scientific American Library, 1990.

Boorse, Henry A., and Lloyd Motz, eds. *The World of the Atom*. 2 vols. New York: Basic Books, 1996.

Chaitin, Gregory J. *The Limits of Mathematics*. Singapore: Springer-Verlag, 1998.

Dewdney, A. K. *The New Turing Omnibus*. New York: Computer Science Press, W. H. Freeman, 1993.

Dudley, Underwood. *Elementary Number Theory*. New York: W. H. Freeman and Company, 1969.

Gamow, George, and Marvin Stern. *Puzzle-Math*. New York: The Viking Press, 1958.

Gardner, Martin. *The Colossal Book of Mathematics*. New York: W. W. Norton & Company, 2001.

Gullberg, Jan. *Mathematics: From the Birth of Numbers*. New York: W. W. Norton & Company, 1997.

Hildebrandt, Stefan, and Anthony Tromba. *Mathematics and Optimal Form*. New York: Scientific American Library, 1985.

Honsberger, Ross. *Mathematical Gems II*. Washington, D.C.: The Mathematical Association of America, 1976.

——. *Mathematical Morsels*. Washington, D.C.: The Mathematical Association of America, 1978.

Honsberger, Ross, ed., *Mathematical Plums*. Washington, D.C.: The Mathematical Association of America, 1979.

Lakatos, Imre. *Proof and Refutations: The Logic of Mathematical Discovery*. Cambridge: Cambridge University Press, 1976.

Morrison, Philip, Phylis Morrison, and the Office of Charles and Rey Eames. *Powers of Ten*. New York: Scientific American Library, 1982.

Newman, James R., ed., *The World of Mathematics*. 4 vols. New York: Simon and Schuster, 1956.

Rózsa, Péter. *Playing with Infinity*. New York: Dover Publications, 1961.

Ulam, S. M. *Adventures of a Mathematician*. New York: Charles Scribner's Sons, 1976.

Weyl, Hermann. *Symmetry*. Princeton, NJ: Princeton University Press, 1952.

· I N D E X ·